N.M. Ferrers

An elementary treatise on spherical harmonics and subjects connected with them

N.M. Ferrers

An elementary treatise on spherical harmonics and subjects connected with them

ISBN/EAN: 9783742891594

Manufactured in Europe, USA, Canada, Australia, Japa

Cover: Foto ©berggeist007 / pixelio.de

Manufactured and distributed by brebook publishing software
(www.brebook.com)

N.M. Ferrers

An elementary treatise on spherical harmonics and subjects connected with them

AN ELEMENTARY TREATISE

ON

SPHERICAL HARMONICS

AND SUBJECTS CONNECTED WITH THEM.

BY THE

REV. N. M. FERRERS, M.A., F.R.S.,

FELLOW AND TUTOR OF GONVILLE AND CAIUS COLLEGE, CAMBRIDGE.

London:

MACMILLAN AND CO.

1877

PREFACE.

The object of the following treatise is to exhibit, in a concise form, the elementary properties of the expressions known by the name of Laplace's Functions, or Spherical Harmonics, and of some other expressions of a similar nature. I do not, of course, profess to have produced a complete treatise on these functions, but merely to have given such an introductory sketch as may facilitate the study of the numerous works and memoirs in which they are employed. As Spherical Harmonics derive their chief interest and utility from their physical applications, I have endeavoured from the outset to keep these applications in view.

I must express my acknowledgments to the Rev. C. H. Prior, Fellow of Pembroke College, for his kind revision of the proof-sheets as they passed through the press.

N. M. FERRERS.

Gonville and Caius College,
August, 1877.

CONTENTS.

CHAPTER I.

INTRODUCTORY. DEFINITION OF SPHERICAL HARMONICS.

CHAPTER II.

ZONAL HARMONICS.

CHAPTER III.

APPLICATION OF ZONAL HARMONICS TO THE THEORY OF ATTRACTION. REPRESENTATION OF DISCONTINUOUS FUNCTIONS BY SERIES OF ZONAL HARMONICS.

CHAPTER IV.

SPHERICAL HARMONICS IN GENERAL. TESSERAL AND SECTORIAL HARMONICS. ZONAL HARMONICS WITH THEIR AXES IN ANY POSITION. POTENTIAL OF A SOLID NEARLY SPHERICAL IN FORM.

CHAPTER V.

SPHERICAL HARMONICS OF THE SECOND KIND.

CHAPTER VI.

ELLIPSOIDAL AND SPHEROIDAL HARMONICS.

ERRATA.

p. 17 line 4, *for* $\frac{1}{2i+1}$, *read* $\frac{2}{2i+1}$.

p. 113 line 8, *for* V *read* E.

p. 136 line 11, *for* ϕ *read* ϖ.

p. 142 line 6, *for* point *read* axis.

CHAPTER I.

INTRODUCTORY. DEFINITION OF SPHERICAL HARMONICS.

1. If V be the potential of an attracting mass, at any point x, y, z, not forming a part of the mass itself, it is known that V must satisfy the differential equation

$$\frac{d^2V}{dx^2}+\frac{d^2V}{dy^2}+\frac{d^2V}{dz^2}=0 \qquad (1),$$

or, as we shall write it for shortness, $\nabla^2 V=0$.

The general solution of this equation cannot be obtained in finite terms. We can, however, determine an expression which we shall call V_i, an homogeneous function of x, y, z of the degree i, i being any positive integer, which will satisfy the equation; and we may prove that to every such solution V_i there corresponds another, of the degree $-(i+1)$, expressed by $\frac{V_i}{r^{2i+1}}$, where $r^2=x^2+y^2+z^2$.

For the equation (1) when transformed to polar co-ordinates by writing $x=r\sin\theta\cos\phi$, $y=r\sin\theta\sin\phi$, $z=r\cos\theta$, becomes

$$r\frac{d^2(rV)}{dr^2}+\frac{1}{\sin\theta}\frac{d}{d\theta}\left(\sin\theta\frac{dV}{d\theta}\right)+\frac{1}{\sin^2\theta}\frac{d^2V}{d\phi^2}=0 \ldots (2).$$

And since V satisfies this equation, and is an homogeneous function of the degree i, V_i must satisfy the equation

$$i(i+1)V_i+\frac{1}{\sin\theta}\frac{d}{d\theta}\left(\sin\theta\frac{dV_i}{d\theta}\right)+\frac{1}{\sin^2\theta}\frac{d^2V_i}{d\phi^2}=0,$$

since this is the form which equation (2) assumes when V is an homogeneous function of the degree i.

Now, put $V_i = r^{2i+1} U_i$, and this becomes

$$i(i+1) r^{2i+1} U_i + \frac{1}{\sin\theta} \frac{d}{d\theta}\left(r^{2i+1} \sin\theta \frac{dU_i}{d\theta}\right) + \frac{1}{\sin^2\theta} \frac{d^2}{d\phi^2}(r^{2i+1} U_i) = 0,$$

or

$$i(i+1) U_i + \frac{1}{\sin\theta} \frac{d}{d\theta}\left(\sin\theta \frac{dU_i}{d\theta}\right) + \frac{1}{\sin^2\theta} \frac{d^2 U_i}{d\phi^2} = 0 \ldots\ldots (2).$$

Now, since U_i is a homogeneous function of the degree $-(i+1)$,

$$\frac{d(rU_i)}{dr} = U_i + r\frac{dU_i}{dr}$$

$$= -iU_i;$$

$$\frac{d^2(rU_i)}{dr^2} = -i\frac{dU_i}{dr}$$

$$= i(i+1)\frac{U_i}{r};$$

or $$r\frac{d^2(rU_i)}{dr^2} = i(i+1) U_i:$$

therefore equation (2) becomes

$$r\frac{d^2(rU_i)}{dr^2} + \frac{1}{\sin\theta} \frac{d}{d\theta}\left(\sin\theta \frac{dU_i}{d\theta}\right) + \frac{1}{\sin^2\theta} \frac{d^2 U_i}{d\phi^2} = 0,$$

shewing that U_i is an admissible value of V, as satisfying equation (2).

It appears therefore that every form of U_i can be obtained from V_i, by dividing by r^{2i+1}, and conversely, that every form of V_i can be obtained from U_i by multiplying by r^{2i+1}. Such an expression as V_i we shall call a Solid Spherical Harmonic of the degree i. The result obtained by dividing V_i by r^i, which will be a function of two independent variables θ and ϕ only, we shall call a Surface Spherical Harmonic of the same degree. A very important form of spherical harmonics is that which is independent of ϕ. The solid harmonics of this form will involve two of the variables, x and y, only in the form x^2+y^2, or will be functions of x^2+y^2 and z. Harmonics independent of ϕ are called Zonal Harmonics, and are distinguished, like spherical harmonics in general, into Solid and Surface Harmonics. The investigation of their properties will be the subject of the following chapter.

The name of Spherical Harmonics was first applied to these functions by Sir W. Thomson and Professor Tait, in their Treatise on Natural Philosophy. The name "Laplace's Coefficients" was employed by Whewell, on account of Laplace having discussed their properties, and employed them largely in the *Mécanique Céleste.* Pratt, in his Treatise on the Figure of the Earth, limits the name of Laplace's *Coefficients* to Zonal Harmonics, and designates all other spherical harmonics by the name of Laplace's *Functions.* The Zonal Harmonic in the case which we shall consider in the following chapter, i.e., in which the system is symmetrical about the line from which θ is measured, was really, however, first introduced by Legendre, although the properties of spherical harmonics in general were first discussed by Laplace; and Mr Todhunter, in his Treatise, on this account calls them by the name of "Legendre's Coefficients," applying the name of "Laplace's Coefficients" to the form which the Zonal Harmonic assumes when in place of $\cos\theta$, we write $\cos\theta\cos\theta'+\sin\theta\sin\theta'\cos(\phi-\phi')$. The name "Kugelfunctionen" is employed by Heine, in his standard treatise on these functions, to designate Spherical Harmonics in general.

CHAPTER II.

ZONAL HARMONICS.

1. WE shall in this chapter regard a Zonal Solid Harmonic, of the degree i, as a homogeneous function of $(x^2+y^2)^{\frac{1}{2}}$, and z, of the degree i, which satisfies the equation

$$\frac{d^2V}{dx^2}+\frac{d^2V}{dy^2}+\frac{d^2V}{dz^2}=0.$$

Now, if this be transformed to polar co-ordinates, by writing $r\sin\theta\cos\phi$ for x, $r\sin\theta\sin\phi$ for y, $r\cos\theta$ for z, we observe, in the first place, that $x^2+y^2=r^2\sin^2\theta$. Hence V will be independent of ϕ, or will be a function of r and θ only. The differential equation between r and θ which it must therefore satisfy will be

$$r\frac{d^2(rV)}{dr^2}+\frac{1}{\sin\theta}\frac{d}{d\theta}\left(\sin\theta\frac{dV}{d\theta}\right)=0.$$

Now V, being a function of r of the degree i, may be expressed in the form r^iP_i, where P_i is a function of θ only. Hence this equation becomes

$$\frac{1}{\sin\theta}\frac{d}{d\theta}\left(\sin\theta\frac{dP_i}{d\theta}\right)+i(i+1)P_i=0, \quad\text{.........(1)},$$

or, putting $\cos\theta=\mu$,

$$\frac{d}{d\mu}\left\{(1-\mu^2)\frac{dP_i}{d\mu}\right\}+i(i+1)P_i=0 \quad\text{......... (2)}.$$

In accordance with our definition of spherical surface harmonics, P_i will be the zonal surface harmonic of the degree i. When it is necessary to particularise the variable involved in it, we shall write it $P_i(\mu)$.

The line from which θ is measured, or in other words for which $\mu=1$, is called the Axis of the system of Zonal Harmonics; and the point in which the positive direction of the axis meets a sphere whose centre is the origin of co-ordinates, and radius unity, is called the Pole of the system.

Any constant multiple of a zonal harmonic (solid or surface) is itself a zonal harmonic of the same order.

2. The zonal harmonic of the degree i, of which the line $\mu=1$ is the axis, is a perfectly determinate function of μ, having nothing arbitrary but this constant. For the expression $r^i P_i$ may be expressed as a rational integral homogeneous function of r and z, and therefore P_i will be a rational integral function of $\cos\theta$, that is of μ, of the degree i, and will involve none but positive integral powers of μ.

But P_i is a particular integral of the equation

$$\frac{d}{d\mu}\left\{(1-\mu^2)\frac{d.f(\mu)}{d\mu}\right\}+i(i+1)f(\mu)=0\text{......(3)},$$

and the most general form of $f(\mu)$ must involve two arbitrary constants. Suppose then that the most general form of $f(\mu)$ is represented by $P_i\int v d\mu$. We then have

$$(1-\mu^2)\frac{d.f(\mu)}{d\mu}=(1-\mu^2)\frac{dP_i}{d\mu}.\int v d\mu+(1-\mu^2)P_i v,$$

$$\frac{d}{d\mu}\left\{(1-\mu^2)\frac{d.f(\mu)}{d\mu}\right\}=\frac{d}{d\mu}\left\{(1-\mu^2)\frac{dP_i}{d\mu}\right\}\int v d\mu$$

$$+2(1-\mu^2)\frac{dP_i}{d\mu}v+P_i\frac{d}{d\mu}\left\{(1-\mu^2).v\right\}.$$

Hence, adding these two equations together, and observing that, since P_i satisfies the equation (3), the coefficient of $\int v d\mu$ will be identically equal to 0, we obtain, for the determination of v, the equation

$$P_i \frac{d}{d\mu}\left\{(1-\mu^2)\,v\right\} + 2(1-\mu^2)\frac{dP_i}{d\mu}\,v = 0,$$

whence $$P_i(1-\mu^2)\frac{dv}{d\mu} + 2\left\{(1-\mu^2)\frac{dP_i}{d\mu} - \mu P_i\right\}v = 0,$$

or $$\frac{dv}{v} + \left(\frac{2}{P_i}\frac{dP_i}{d\mu} - \frac{2\mu}{1-\mu^2}\right)d\mu = 0,$$

the integral of which is

$$\log v + \log P_i^2(1-\mu^2) = \log C_1 = \text{a constant};$$

$$\therefore\ v = \frac{C_1}{P_i^2(1-\mu^2)}.$$

Hence $$\int v d\mu = C + C_1\int\frac{d\mu}{P_i^2(1-\mu^2)};$$

and we obtain, for the most general form of $f(\mu)$,

$$f(\mu) = CP_i + C_1 P_i\int\frac{d\mu}{P_i^2(1-\mu^2)}.$$

Now, P_i being a rational integral function of μ of i dimensions, it may be seen that $\int\frac{d\mu}{(1-\mu^2)P_i^2}$ will assume the form of the sum of $i+2$ logarithms and i fractions, and therefore cannot be expressed as a rational integral function of μ. Expressions of the form $P_i\int\frac{d\mu}{(1-\mu^2)P_i^2}$ are called *Kugelfunctionen der zweiter Art* by Heine, who has investigated their properties at great length. They have, as will hereafter be seen, interesting applications to the attraction of a spheroid on an external point. We shall discuss their properties more fully hereafter.

3. We have thus shewn that the most general solution of equation (2) of the form of a rational integral function of μ

involves but one arbitrary constant, and that as a factor. We shall henceforth denote by P_i, or $P_i(\mu)$, that particular form of the integral which assumes the value unity when μ is put equal to unity.

We shall next prove the following important proposition.

If h *be less than unity, and if* $(1 - 2\mu h + h^2)^{-\frac{1}{2}}$ *be expanded in a series proceeding by ascending powers of* h, *the coefficient of* h^i *will be* P_i.

Or, $$(1 - 2\mu h + h^2)^{-\frac{1}{2}} = P_0 + P_1 h + \ldots + P_i h^i + \ldots$$

We shall prove this by shewing that, if H be written for $(1 - 2\mu h + h^2)^{-\frac{1}{2}}$, H will satisfy the differential equation

$$\frac{d}{d\mu}\left\{(1-\mu^2)\frac{dH}{d\mu}\right\} + h\frac{d^2}{dh^2}(hH) = 0.$$

For, since $$H = (1 - 2\mu h + h^2)^{-\frac{1}{2}},$$

$$\therefore \frac{1}{H^2} = 1 - 2\mu h + h^2;$$

$$\therefore \frac{1}{H^3}\frac{dH}{d\mu} = h,$$

$$\text{or } \frac{1}{h}\frac{dH}{d\mu} = H^3;$$

$$\therefore \frac{1}{h}\frac{d}{d\mu}\left\{(1-\mu^2)\frac{dH}{d\mu}\right\} = \frac{d}{d\mu}\left\{(1-\mu^2)H^3\right\}$$

$$= -2\mu H^3 + 3(1-\mu^2)H^2\frac{dH}{d\mu}$$

$$= -2\mu H^3 + 3(1-\mu^2)hH^5.$$

And $$\frac{1}{H^3}\frac{dH}{dh} = \mu - h,$$

$$\therefore \frac{d}{dh}(hH) = H + h\frac{dH}{dh} = H^3\left(\frac{1}{H^2} + \frac{h}{H^3}\frac{dH}{dh}\right)$$

$$= H^3\{1 - 2\mu h + h^2 + h(\mu - h)\}$$

$$= H^3(1 - \mu h);$$

$$\therefore \frac{d^2}{dh^2}(hH) = \frac{d}{dh}\{H^3(1-\mu h)\}$$

$$= 3(1-\mu h)H^2\frac{dH}{dh} - \mu H^3$$

$$= 3(1-\mu h)H^5(\mu - h) - \mu H^3.$$

Hence $$\frac{1}{h}\frac{d}{d\mu}\left\{(1-\mu^2)\frac{dH}{d\mu}\right\} + \frac{d^2}{dh^2}(hH)$$

$$= -3\mu H^3 + 3\{(1-\mu^2)h + (1-\mu h)(\mu - h)\}H^5$$

$$= -3\mu H^3 + 3\{\mu(1+h^2) - 2\mu^2 h\}H^5$$

$$= -3\mu\{H^3 - (1-2\mu h + h^2)H^5\}$$

$$= 0, \text{ since } 1 - 2\mu h + h^2 = H^{-2}.$$

Therefore $$\frac{d}{d\mu}\left\{(1-\mu^2)\frac{dH}{d\mu}\right\} + h\frac{d^2}{dh^2}(hH) = 0.$$

This may also be shewn as follows.

If x, y, z be the co-ordinates of any point, z' the distance of a fixed point, situated on the axis of z, from the origin, and R be the distance between these points, we know that,

$$R^2 = x^2 + y^2 + (z' - z)^2,$$

and that $$\nabla^2\left(\frac{1}{R}\right) = 0.$$

Now, transform these expressions to polar co-ordinates, by writing

$$x = r\sin\theta\cos\phi, \quad y = r\sin\theta\sin\phi, \quad z = r\cos\theta,$$

and we get

$$R^2 = r^2 - 2z'r\cos\theta + z'^2,$$

and the differential equation becomes

$$r\frac{d^2}{dr^2}\left(\frac{r}{R}\right) + \frac{1}{\sin\theta}\frac{d}{d\theta}\left\{\sin\theta\frac{d}{d\theta}\left(\frac{1}{R}\right)\right\} = 0,$$

or, putting $\cos\theta = \mu$,

$$r\frac{d^2}{dr^2}\left(\frac{r}{R}\right) + \frac{d}{d\mu}\left\{(1-\mu^2)\frac{d}{d\mu}\left(\frac{1}{R}\right)\right\} = 0.$$

Now, putting $r = z'h$, we see that

$$\frac{R^2}{z'^2} = h^2 - 2\mu h + 1 = \frac{1}{H^2},$$

or

$$\frac{1}{R} = \frac{H}{z'},$$

$$\therefore \frac{r}{R} = hH,$$

and

$$r\frac{d^2}{dr^2}\left(\frac{r}{R}\right) = \frac{h}{z'}\frac{d^2}{dh^2}(hH).$$

$\therefore$ the above equation becomes

$$\frac{h}{z'}\frac{d^2}{dh^2}(hH) + \frac{d}{d\mu}\left\{(1-\mu^2)\frac{d}{d\mu}\left(\frac{H}{z'}\right)\right\} = 0,$$

or

$$h\frac{d^2(hH)}{dh^2} + \frac{d}{d\mu}\left\{(1-\mu^2)\frac{dH}{d\mu}\right\} = 0.$$

4. Having established this proposition, we may proceed as follows:

If p_i be the coefficient of h^i in the expansion of H,

$$H = 1 + p_1 h + p_2 h^2 + \ldots + p_i h^i + \ldots$$

$$\therefore hH = h + p_1 h^2 + p_2 h^3 + \ldots + p_i h^{i+1} + \ldots$$

$$\therefore h\frac{d^2}{dh^2}(hH) = 1 \,.\, 2 p_1 h + 2 \,.\, 3 p_2 h^2 + \ldots + i(i+1) p_i h^i + \ldots$$

Also, the coefficient of h^i in the expansion of

$$\frac{d}{d\mu}\left\{(1-\mu^2)\frac{dH}{d\mu}\right\} \text{ is } \frac{d}{d\mu}\left\{(1-\mu^2)\frac{dp_i}{d\mu}\right\}.$$

Hence equating to zero the coefficient of h^i,

$$\frac{d}{d\mu}\left\{(1-\mu^2)\frac{dp_i}{d\mu}\right\} + i(i+1)p_i = 0.$$

Also p_i is a rational integral function of μ.

And, when $\mu = 1$, $H = (1 - 2h + h^2)^{-\frac{1}{2}}$

$$= 1 + h + h^2 + \ldots + h^i + \ldots$$

Or when $\mu = 1$, $p_i = 1$.

Therefore p_i is what we have already denoted by P_i.

We have thus shewn that, if h be less than 1,

$$(1 - 2\mu h + h^2)^{-\frac{1}{2}} = P_0 + P_1 h + \ldots + P_i h^i + \ldots$$

If h be greater than 1, this series becomes divergent.

But we may write

$$(h^2 - 2\mu h + 1)^{-\frac{1}{2}} = \frac{1}{h}\left(1 - 2\frac{\mu}{h} + \frac{1}{h^2}\right)^{-\frac{1}{2}}$$

$$= \frac{1}{h}\left(P_0 + \frac{P_1}{h} + \ldots + \frac{P_i}{h^i} + \ldots\right),$$

since $\frac{1}{h}$ is less than 1,

$$= \frac{P_0}{h} + \frac{P_1}{h^2} + \ldots + \frac{P_i}{h^{i+1}} + \ldots.$$

Hence P_i is also the coefficient of $h^{-(i+1)}$ in the expansion of $(1 - 2\mu h + h^2)^{-\frac{1}{2}}$ in ascending powers of $\frac{1}{h}$ when h is greater than 1. We may express this in a notation which is strictly continuous, by saying that

$$P_i = P_{-(i+1)}.$$

This might have been anticipated, from the fact that the fundamental differential equation for P_i is unaltered if $-(i+1)$ be written in place of i; for the only way in which i appears in that equation is in the coefficient of P_i, which is $i(i+1)$. Writing $-(i+1)$ in place of i, this becomes $-(i+1)\{-(i+1)+1\}$ or $(i+1)i$, and is therefore unaltered.

5. We shall next prove that

$$P_i = (-1)^i \frac{r^{i+1}}{1.2\ldots i} \frac{d^i}{dz^i}\left(\frac{1}{r}\right),$$

where $r^2 = x^2 + y^2 + z^2$.

Let $$\frac{1}{r} = (x^2 + y^2 + z^2)^{-\frac{1}{2}} = f(z),$$

and let k be any quantity less than r.

Then $$\{x^2 + y^2 + (z-k)^2\}^{-\frac{1}{2}} = f(z-k),$$

and, developing by Taylor's Theorem, the coefficient of k^i is

$$(-1)^i \frac{f^i(z)}{1.2\ldots i}, \text{ or } (-1)^i \frac{1}{1.2\ldots i} \frac{d^i}{dz^i}\left(\frac{1}{r}\right).$$

Also $$\{x^2 + y^2 + (z-k)^2\}^{-\frac{1}{2}} = (r^2 - 2kz + k^2)^{-\frac{1}{2}}$$

$$= \frac{1}{r}\left(1 - 2\mu\frac{k}{r} + \frac{k^2}{r^2}\right)^{-\frac{1}{2}},$$

since $z = \mu r$,

in the expansion of which, the coefficient of k^i is

$$\frac{P_i}{r^{i+1}}.$$

Equating these results, we get

$$P_i = (-1)^i \frac{r^{i+1}}{1.2\ldots i} \frac{d^i}{dz^i}\left(\frac{1}{r}\right).$$

The value of P_i might be calculated, either by expanding $(1 - 2\mu h + h^2)^{-\frac{1}{2}}$ by the Binomial Theorem, or by effecting the differentiations in the expression $(-1)^i \frac{r^{i+1}}{1.2.3\ldots i} \frac{d^i}{dz^i}\left(\frac{1}{r}\right)$, and in the result putting $\frac{z}{r} = \mu$. Both these methods however would be somewhat laborious; we proceed therefore to investigate more convenient expressions.

6. The first process shews, by the aid of Lagrange's Theorem, that

$$P_i = \frac{1}{2^i . 1 . 2 . 3 \dots i} \frac{d^i}{d\mu^i} (\mu^2 - 1)^i.$$

Let y denote a quantity, such that

$$y = \frac{1}{h} - \left(1 - \frac{2\mu}{h} + \frac{1}{h^2}\right)^{\frac{1}{2}},$$

h being less than 1.

Then

$$\frac{dy}{d\mu} = \frac{\frac{1}{h}}{\left(1 - \frac{2\mu}{h} + \frac{1}{h^2}\right)^{\frac{1}{2}}} = \frac{1}{(1 - 2\mu h + h^2)^{\frac{1}{2}}}.$$

Also
$$\left(y - \frac{1}{h}\right)^2 = 1 - \frac{2\mu}{h} + \frac{1}{h^2};$$

$$\therefore\ y^2 - \frac{2y}{h} = 1 - \frac{2\mu}{h};$$

$$\therefore\ y = \mu + h\left(\frac{y^2 - 1}{2}\right).$$

Hence, by Lagrange's Theorem,

$$y = \mu + h\frac{\mu^2 - 1}{2} + \frac{h^2}{1 . 2}\frac{d}{d\mu}\left(\frac{\mu^2 - 1}{2}\right)^2 + \dots$$
$$+ \frac{h^i}{1 . 2 \dots i}\frac{d^{i-1}}{d\mu^{i-1}}\left(\frac{\mu^2 - 1}{2}\right)^i + \dots,$$

therefore, differentiating with respect to μ and observing that

$$\frac{dy}{d\mu} = (1 - 2\mu h + h^2)^{-\frac{1}{2}},$$

$$(1 - 2\mu h + h^2)^{-\frac{1}{2}} = 1 + h\frac{d}{d\mu}\left(\frac{\mu^2 - 1}{2}\right) + \frac{h^2}{1 . 2}\frac{d^2}{d\mu^2}\left(\frac{\mu^2 - 1}{2}\right)^2 + \dots$$
$$+ \frac{h^i}{1 . 2 \dots i}\frac{d^i}{d\mu^i}\left(\frac{\mu^2 - 1}{2}\right)^i + \dots.$$

Hence $$P_i = \frac{1}{2^i . 1 . 2 \ldots i} \frac{d^i}{d\mu^i} (\mu^2 - 1)^i.$$

7. From this form of P_i it may be readily shewn that the values of μ, which satisfy the equation $P_i = 0$, are all real, and all lie between -1 and 1.

For the equation

$(\mu^2 - 1)^i = 0$ has i roots $= 1$, and i roots $= -1$,

$\therefore \frac{d}{d\mu} (\mu^2 - 1)^i = 0$ has $i - 1$ roots $= 1$, $(i - 1)$ roots $= -1$, and one root $= 0$,

$\frac{d^2}{d\mu^2} (\mu^2 - 1)^i = 0$ has $(i - 2)$ roots $= 1$, one root between 1 and 0, one between 0 and $= -1$, and $(i - 2)$ roots $= -1$, and so on. Hence it follows that

$\frac{d^i}{d\mu^i} (\mu^2 - 1)^i = 0$ has $\frac{i}{2}$ roots between 1 and 0, and $\frac{i}{2}$ roots between 0 and -1, if i be even,

and $\frac{i-1}{2}$ roots between 1 and 0, $\frac{i-1}{2}$ roots between 0 and 1, and one root $= 0$, if i be odd.

It is hardly necessary to observe that the positive roots of each of these equations are severally equal in absolute magnitude to the negative roots.

8. We may take this opportunity of introducing an important theorem, due to Rodrigues, properly belonging to the Differential Calculus, but which is of great use in this subject.

The theorem in question is as follows:

If m *be any integer less than* i,

$$\frac{d^{i-m}}{dx^{i-m}} (x^2 - 1)^i = \frac{1 . 2 \ldots (i - m)}{1 . 2 \ldots (i + m)} (x^2 - 1)^m \frac{d^{i+m}}{dx^{i+m}} (x^2 - 1)^i.$$

It may be proved in the following manner.

If $(x^2-1)^i$ be differentiated $i-m$ times, then, since the equation

$$(x^2-1)^i=0$$

has i roots each equal to 1, and i roots each equal $=-1$, it follows that the equation

$$\frac{d^{i-m}}{dx^{i-m}}(x-1)^i=0$$

has $i-(i-m)$ roots (i. e. m) roots each $=1$, and m roots each $=-1$, in other words that $(x^2-1)^m$ is a factor of

$$\frac{d^{i-m}}{dx^{i-m}}(x^2-1)^i.$$

We proceed to calculate the other factor.

For this purpose consider the expression

$$(x+\alpha_1)(x+\alpha_2)\ldots(x+\alpha_i)(x+\beta_1)(x+\beta_2)\ldots(x+\beta_i).$$

Conceive this differentiated (I) $i-m$ times, (II) $i+m$ times. The two expressions thus obtained will consist of an equal number of terms, and to any term in (I) will correspond one term in (II), such that their product will be $(x+\alpha_1)(x+\alpha_2)\ldots(x+\alpha_i)(x+\beta_1)(x+\beta_2)\ldots(x+\beta_i)$, i.e. the term in (II) is the product of all the factors omitted from the corresponding term in (I) and of those factors only. Two such terms may be said to be complementary to each other.

Now, conceive a term in (II) the product of p factors of the form $x+\alpha$, say $x+\alpha'$, $x+\alpha''\ldots x+\alpha^{(p)}$, and of q factors of the form $x+\beta$, say $x+\beta_{,}$, $x+\beta_{,,}\ldots x+\beta_{(q)}$. We must have $p+q=i-m$.

The complementary term in (I) will involve

$$p \text{ factors } x+\beta',\ x+\beta''\ \ldots\ x+\beta^{(p)},$$

$$q \text{ factors } x+\alpha_{,},\ x+\alpha_{,,}\ \ldots\ x+\alpha_{(q)}.$$

Now, every term in (I) is of $i+m$ dimensions. We have accounted for $p+q$ (or $i-m$) factors in the particular term we are considering. There remain therefore $2m$ factors to be accounted for. None of the letters

$$\alpha', \alpha'' \ldots \alpha^{(p)}, \beta_{,}, \beta_{,,} \ldots \beta_{(q)},$$
$$\beta', \beta'' \ldots \beta^{(p)}, \alpha_{,}, \alpha_{,,} \ldots \alpha_{(q)},$$

can appear there. Hence the remaining factor must involve m α's and m β's,—say,

$$_1\alpha, {}_2\alpha \ldots {}_m\alpha,$$
$$_1\beta, {}_2\beta \ldots {}_m\beta.$$

There will be another term in (II) containing

$$(x+\beta')(x+\beta'') \ldots (x+\beta^{(p)})(x+\alpha_{,})(x+\alpha_{,,}) \ldots (x+\alpha_{(q)}).$$

The corresponding term in (I) will be, as shewn above,

$$(x+\alpha')(x+\alpha'') \ldots (x+\alpha^{(p)})(x+\beta_{,})(x+\beta_{,,}) \ldots (x+\beta_{(q)})$$
$$(x+{}_1\alpha)(x+{}_2\alpha) \ldots (x+{}_m\alpha)(x+{}_1\beta)(x+{}_2\beta) \ldots (x+{}_m\beta).$$

Hence, the sum of these two terms of (I) divided by the sum of the complementary two terms of (II) is

$$(x+{}_1\alpha)(x+{}_2\alpha) \ldots (x+{}_m\alpha)(x+{}_1\beta)(x+{}_2\beta) \ldots (x+{}_m\beta).$$

Now, let each of the α's be equal to 1, and each of the β's equal to -1, then this becomes $(x^2-1)^m$. The same factor enters into every such pair of the terms of (I). Hence

$$\frac{(\text{I})}{(\text{II})} = (x^2-1)^m.$$

Or $\dfrac{d^{i-m}(x^2-1)^i}{dx^{i-m}} = (x^2-1)^m \dfrac{d^{i+m}(x^2-1)^i}{dx^{i+m}}$, to a numerical factor *près*.

The factor may easily be calculated, by considering that the coefficient of x^{i+m} in $\dfrac{d^{i-m}(x^2-1)^i}{dx^{i-m}}$ is $2i(2i-1)\ldots(i+m+1)$, and that the coefficient of x^{i-m} in $\dfrac{d^{i+m}(x^2-1)^i}{dx^{i+m}}$ is

$$2i(2i-1)\ldots(i+m+1)(i+m)\ldots(i-m+1).$$

Hence the factor is

$$\frac{1}{(i+m)(i+m-1)\ldots(i-m+1)}, \text{ or } \frac{1.2\ldots(i-m)}{1.2\ldots(i+m)}.$$

9. This theorem affords a direct proof that $C\frac{d^i}{d\mu^i}(\mu^2-1)^i$, C being any constant, is a value of $f(\mu)$ which satisfies the equation

$$\frac{d}{d\mu}\left\{(1-\mu^2)\frac{df(\mu)}{d\mu}\right\}+i(i+1)f(\mu)=0.$$

For $(\mu^2-1)\frac{d}{d\mu}\cdot\frac{d^i}{d\mu^i}(\mu^2-1)^i=(\mu^2-1)\frac{d^{i+1}}{d\mu^{i+1}}(\mu^2-1)^i$

$$=i(i+1)\frac{d^{i-1}}{d\mu^{i-1}}(\mu^2-1)^i$$

from above,

$$\therefore \frac{d}{d\mu}\left[(\mu^2-1)\frac{d}{d\mu}\left\{\frac{d^i}{d\mu^i}(\mu^2-1)^i\right\}\right]=i(i+1)\left\{\frac{d^i}{d\mu^i}(\mu^2-1)^i\right\},$$

or

$$\frac{d}{d\mu}\left[(1-\mu^2)\frac{d}{d\mu}\left\{\frac{d^i}{d\mu^i}(\mu^2-1)^i\right\}\right]+i(i+1)\left\{\frac{d^i}{d\mu^i}(\mu^2-1)^i\right\}=0.$$

Hence, the given differential equation is satisfied by putting $f(\mu)=C\frac{d^i}{d\mu^i}(\mu^2-1)^i$.

Introducing the condition that P_i is that value of $f(\mu)$ which is equal to 1, when $\mu=1$, we get

$$P_i=\frac{1}{2^i.1.2\ldots i}\frac{d^i}{d\mu^i}(\mu^2-1)^i.$$

10. We shall now establish two very important properties of the function P_i; and apply them to obtain the development of P_i in a series.

The properties in question are as follows:

If i *and* m *be unequal positive integers,*

$$\int_{-1}^{1} P_i P_m d\mu = 0.$$

And
$$\int_{-1}^{1} P_i^2 d\mu = \frac{1}{2i+1}.$$

The following is a proof of the first property.

We have

$$\frac{d}{d\mu}\left\{(1-\mu^2)\frac{dP_i}{d\mu}\right\} + i(i+1)P_i = 0,$$

$$\frac{d}{d\mu}\left\{(1-\mu^2)\frac{dP_m}{d\mu}\right\} + m(m+1)P_m = 0.$$

Multiplying the first of these equations by P_m, the second by P_i, subtracting and integrating, we get

$$\int P_m \frac{d}{d\mu}\left\{(1-\mu^2)\frac{dP_i}{d\mu}\right\} d\mu - \int P_i \frac{d}{d\mu}\left\{(1-\mu^2)\frac{dP_m}{d\mu}\right\} d\mu$$
$$+ \{i(i+1) - m(m+1)\} \int P_i P_m d\mu = 0.$$

Hence, transforming the first two integrals by integration by parts, and remarking that

$$i(i+1) - m(m+1) = (i-m)(i+m+1),$$

we get

$$(1-\mu^2)\left(P_m \frac{dP_i}{d\mu} - P_i \frac{dP_m}{d\mu}\right) - \int (1-\mu^2)\left(\frac{dP_m}{d\mu}\frac{dP_i}{d\mu} - \frac{dP_i}{d\mu}\frac{dP_m}{d\mu}\right) d\mu$$
$$+ (i-m)(i+m+1)\int P_i P_m d\mu = 0,$$

or

$$(1-\mu^2)\left(P_m \frac{dP_i}{d\mu} - P_i \frac{dP_m}{d\mu}\right) + (i-m)(i+m+1)\int P_i P_m d\mu = 0,$$

since the second term vanishes identically.

Hence, taking the integral between the limits -1 and $+1$, we remark that the factor $1-\mu^2$ vanishes at both limits, and therefore, *except when* $i-m$, *or* $i+m+1=0$,

$$\int_{-1}^{1} P_i P_m d\mu = 0.$$

We may remark also that we have in general

$$\int_{\mu}^{1} P_i P_m d\mu = (1-\mu^2)\frac{P_m \frac{dP_i}{d\mu} - P_i \frac{dP_m}{d\mu}}{(i-m)(i+m+1)},$$

a result which will be useful hereafter.

11. We will now consider the cases in which

$$i-m, \text{ or } i+m+1=0.$$

We see that $i+m+1$ cannot be equal to 0, if i and m are both positive integers. Hence we need only discuss the case in which $m=i$. We may remark, however, that since $P_i = P_{-(i+1)}$, the determination of the value of $\int_{-1}^{1} P_i^2 d\mu$ will also give the value of $\int_{-1}^{1} P_i P_{-(i+1)} d\mu$.

The value of $\int_{-1}^{1} P_i^2 d\mu$ may be calculated as follows:

$$(1-2\mu h+h^2)^{-\frac{1}{2}} = P_0 + P_1 h + \ldots + P_i h^i + \ldots;$$

$$\begin{aligned}\therefore\ (1-2\mu h+h^2)^{-1} &= (P_0 + P_1 h + \ldots + P_i h^i + \ldots)^2 \\ &= P_0^2 + P_1^2 h^2 + \ldots + P_i^2 h^{2i} + \ldots \\ &+ 2P_0 P_1 h + 2P_0 P_2 h^2 + \ldots + 2P_1 P_2 h^3 + \ldots\end{aligned}$$

Integrate both sides with respect to μ; then since

$$\int (1-2\mu h+h^2)^{-1} d\mu = -\frac{1}{2h}\log(1-2\mu h+h^2),$$

we get, taking this integral between the limits -1 and $+1$,

$$\frac{1}{h}\log\frac{1+h}{1-h} = \int_{-1}^{1} P_0^2 d\mu + h^2\int_{-1}^{1} P_1^2 d\mu + \ldots + h^{2i}\int_{-1}^{1} P_i^2 d\mu + \ldots$$

all the other terms vanishing, by the theorem just proved.

Now $\log \frac{1+h}{1-h} = 2\left(h + \frac{h^3}{3} + \ldots + \frac{h^{2i+1}}{2i+1} + \ldots\right)$.

Hence $2\left(1 + \frac{h^2}{3} + \ldots + \frac{h^{2i}}{2i+1} + \ldots\right)$

$$= \int_{-1}^{1} P_0^2 d\mu + h^2 \int_{-1}^{1} P_1^2 d\mu + \ldots + h^{2i} \int_{-1}^{1} P_i^2 d\mu + \ldots$$

Hence, equating coefficients of h^{2i},

$$\int_{-1}^{1} P_i^2 d\mu = \frac{2}{2i+1}.$$

12. From the equation $\int_{-1}^{1} P_i P_m d\mu = 0$, combined with the fact that, when $\mu = 1$, $P_i = 1$, and that P_i is a rational integral function of μ, of the degree i, P_i may be expressed in a series by the following method.

We may observe in the first place that, if m be any integer less than i, $\int_{-1}^{1} \mu^m P_i d\mu = 0$.

For as P_m, P_{m-1} ... may all be expressed as rational integral functions of μ, of the degrees m, $m-1$ ... respectively, it follows that μ^m will be a linear function of P_m and zonal harmonics of lower orders, μ^{m-1} of P_{m-1} and zonal harmonics of lower orders, and so on. Hence $\int \mu^m P_i d\mu$ will be the sum of a series of multiples of quantities of the form $\int P_m P_i d\mu$, m being less than i, and therefore $\int_{-1}^{1} \mu^m P_i d\mu = 0$, if m be any integer less than i.

Again, since

$$(1 - 2\mu h + h^2)^{-\frac{1}{2}} = P_0 + P_1 h + \ldots + P_i h^i + \ldots$$

it follows, writing $-h$ for h, that

$$(1 + 2\mu h + h^2)^{-\frac{1}{2}} = P_0 - P_1 h + \ldots + (-1)^i P_i h^i + \ldots$$

And writing $-\mu$ for μ in the first equation,

$$(1+2\mu h+h^2)^{-\frac{1}{2}}=P_0'+P_1'h+\ldots+P_i'h^i+\ldots$$

P_0', $P_1' \ldots P_i' \ldots$ denoting the values which P_0, $P_1 \ldots P_i$, respectively assume, when $-\mu$ is written for μ. Hence $P_i'=P_i$ or $-P_i$, according as i is even or odd. That is, P_i involves only odd, or only even, powers of μ, according as i is odd or even*.

Assume then

$$P_i=A_i\mu^i+A_{i-2}\mu^{i-2}+\ldots$$

Our object is to determine A_i, $A_{i-2}\ldots$.

Then, multiplying successively by μ^{i-2}, μ^{i-4}, ... and integrating from -1 to $+1$, we obtain the following system of equations:

$$\frac{A_i}{2i-1}+\frac{A_{i-2}}{2i-3}+\ldots+\frac{A_{i-2s}}{2i-2s-1}+\ldots=0,$$

$$\frac{A_i}{2i-3}+\frac{A_{i-2}}{2i-5}+\ldots+\frac{A_{i-2s}}{2i-2s-3}+\ldots=0,$$

....

$$\frac{A_i}{2i-2s-1}+\frac{A_{i-2}}{2i-2s-3}+\ldots+\frac{A_{i-2s}}{2i-4s-1}+\ldots=0.$$

....

And lastly, since $P_i=1$, when $\mu=1$,

$$A_i+A_{i-2}+\ldots+A_{i-2s}+\ldots=1;$$

the last terms of the first members of these several equations being

$$\frac{A_0}{i-1},\ \frac{A_0}{i-3}\ldots\frac{A_0}{1},\ A_0, \text{ if } i \text{ be even},$$

$$\frac{A_1}{i-2},\ \frac{A_1}{i-4}\ldots\frac{A_1}{2},\ A_1, \text{ if } i \text{ be odd}.$$

13. The mode of solving the class of systems of equations to which this system belongs will be best seen by considering a particular example.

* This is also evident, from the fact that P_i is a constant multiple of $\frac{d^i}{d\mu^i}(\mu^2-1)^i$.

Suppose then that we have

$$\frac{x}{a+\alpha}+\frac{y}{b+\alpha}+\frac{z}{c+\alpha}=0,$$

$$\frac{x}{a+\beta}+\frac{y}{b+\beta}+\frac{z}{c+\beta}=0,$$

$$\frac{x}{a+\omega}+\frac{y}{b+\omega}+\frac{z}{c+\omega}=\frac{1}{\omega}.$$

From this system of equations we deduce the following, θ being any quantity whatever,

$$\frac{x}{a+\theta}+\frac{y}{b+\theta}+\frac{z}{c+\theta}=\frac{1}{\omega}\frac{(\theta-\alpha)(\theta-\beta)(a+\omega)(b+\omega)(c+\omega)}{(\omega-\alpha)(\omega-\beta)(a+\theta)(b+\theta)(c+\theta)}.$$

For this expression is of -1 dimension in $a, b, c, \alpha, \beta, \gamma, \theta, \omega$; it vanishes when $\theta=\alpha$, or $\theta=\beta$, and for no other finite value of θ, and it becomes $=\frac{1}{\omega}$, when $\theta=\omega$.

We hence obtain

$$x+(a+\theta)\left(\frac{y}{b+\theta}+\frac{z}{c+\theta}\right)=\frac{1}{\omega}\frac{(\theta-\alpha)(\theta-\beta)}{(\omega-\alpha)(\omega-\beta)}\frac{(a+\omega)(b+\omega)(c+\omega)}{(b+\theta)(c+\theta)},$$

and therefore, putting $\theta=-a$,

$$x=\frac{1}{\omega}\frac{(a+\alpha)(a+\beta)}{(a-b)(a-c)}\frac{(a+\omega)(b+\omega)(c+\omega)}{(\omega-\alpha)(\omega-\beta)},$$

with similar values for y and z.

And, if ω be infinitely great, in which case the last equation assumes the form $x+y+z=1$, we have

$$x=\frac{(a+\alpha)(a+\beta)}{(a-b)(a-c)},$$

with similar values for y and z.

14. Now consider the general system

$$\frac{x_i}{a_i+\alpha_i}+\frac{x_{i-2}}{a_{i-2}+\alpha_i}+\ldots+\frac{x_{i-2s}}{a_{i-2s}+\alpha_i}+\ldots=0,$$

$$\frac{x_i}{a_i+\alpha_{i-2}}+\frac{x_{i-2}}{a_{i-2}+\alpha_{i-2}}+\ldots+\frac{x_{i-2s}}{a_{i-2s}+\alpha_{i-2}}+\ldots=0,$$

$$\frac{x_i}{a_i+\alpha_{i-2s}}+\frac{x_{i-2}}{a_{i-2}+\alpha_{i-2s}}+\ldots+\frac{x_{i-2s}}{a_{i-2s}+\alpha_{i-2s}}+\ldots=0,$$

....

$$\frac{x_i}{a_i+\omega}+\frac{x_{i-2}}{a_{i-2}+\omega}+\ldots+\frac{x_{i-2s}}{a_{i-2s}+\omega}+\ldots=\frac{1}{\omega};$$

the number of equations, and therefore of letters of the forms x and a, being $\frac{i+1}{2}$ if i be odd, $\frac{i}{2}+1$ if i be even; and the number of letters of the form α being $\frac{i-1}{2}$ if i be odd, and $\frac{i}{2}-1$ if i be even.

We obtain, as before,

$$\frac{x_i}{a_i+\theta}+\frac{x_{i-2}}{a_{i-2}+\theta}+\ldots+\frac{x_{i-2s}}{a_{i-2s}+\theta}+\ldots$$

$$=\frac{1}{\omega}\frac{(\theta-\alpha_i)(\theta-\alpha_{i-2})\ldots(\theta-\alpha_{i-2s})\ldots}{(\omega-\alpha_i)(\omega-\alpha_{i-2})\ldots(\omega-\alpha_{i-2s})\ldots}\frac{(a_i+\omega)(a_{i-2}+\omega)\ldots(a_{i-2s}+\omega)\ldots}{(a_i+\theta)(a_{i-2}+\theta)\ldots(a_{i-2s}+\theta)\ldots}$$

and, multiplying by $a_{i-2s}+\theta$, and then putting $\theta=-a_{i-2s}$,

$$x_{i-2s}=\frac{1}{\omega}\frac{(a_{i-2s}+\alpha_i)(a_{i-2s}+\alpha_{i-2})\ldots(a_{i-2s}+\alpha_{i-2s})\ldots}{(\omega-\alpha_i)(\omega-\alpha_{i-2})\ldots(\omega-\alpha_{i-2s})\ldots}$$

$$\frac{(a_i+\omega)(a_{i-2}+\omega)\ldots(a_{i-2s}+\omega)\ldots}{(a_{i-2s}-a_i)(a_{i-2s}-a_{i-2})\ldots(a_{i-2s}-a_1 \text{ or } a_0)}.$$

15. To apply this to the case of zonal harmonics, we see, by comparing the equations for x with the equations for A, that we must suppose $\omega=\infty$; and

$$a_i=i,\ a_{i-2}=i-2,\ldots a_{i-2s}=i-2s\ldots$$

$$\alpha_i=i-1,\ \alpha_{i-2}=i-3,\ldots\alpha_{i-2s}=i-2s-1\ldots$$

Hence

$$A_{i-2s}=\frac{(2i-2s-1)(2i-2s-3)\ldots\{2(i-2s)-1\}\ldots}{(-2s)\{-(2s-2)\}\ldots\{(i-2s-1)\ \text{ or }\ (i-2s)\}}$$

$$=(-1)^s\frac{(2i-2s-1)(2i-2s-3)\ldots\{2(i-2s)-1\}\ldots}{2s(2s-2)\ldots2\times2\,.\,4\ldots(i-2s-1)\text{ or }(i-2s)}.$$

Or, generally, if i be odd,

$$A_i = \frac{(2i-1)(2i-3)\ldots(i+2)}{2.4\ldots(i-1)},$$

$$A_{i-2} = -\frac{(2i-3)(2i-5)\ldots i}{2.4\ldots(i-3)\times 2},$$

$$A_{i-4} = \frac{(2i-5)(2i-7)\ldots(i-2)}{2.4\ldots(i-5)\times 2.4},$$

$$\ldots\ldots = \ldots\ldots\ldots\ldots\ldots\ldots\ldots\ldots$$

$$A_1 = (-1)^{\frac{i-1}{2}}\frac{i(i-2)\ldots 3}{2.4\ldots(i-1)}.$$

And, if i be even,

$$A_i = \frac{(2i-1)(2i-3)\ldots(i+1)}{2.4\ldots i},$$

$$A_{i-2} = -\frac{(2i-3)(2i-5)\ldots(i-1)}{2.4\ldots(i-2)\times 2},$$

$$A_{i-4} = \frac{(2i-5)(2i-7)\ldots(i-3)}{2.4\ldots(i-4)\times 2.4},$$

$$\ldots\ldots = \ldots\ldots\ldots\ldots\ldots\ldots\ldots\ldots$$

$$A_0 = (-1)^{\frac{i}{2}}\frac{(i-1)(i-3)\ldots 1}{2.4\ldots i}.$$

We give the values of the several zonal harmonics, from P_0 to P_{10} inclusive, calculated by this formula,

$$P_0 = 1,$$

$$P_1 = \mu,$$

$$P_2 = \frac{3}{2}\mu^2 - \frac{1}{2},$$

$$= \frac{3\mu^2 - 1}{2},$$

$$P_3 = \frac{5}{2}\mu^3 - \frac{3}{2}\mu$$

$$= \frac{5\mu^3 - 3\mu}{2},$$

$$P_4 = \frac{7.5}{2.4}\mu^4 - \frac{5.3}{2.2}\mu^2 + \frac{3.1}{2.4}$$

$$= \frac{35\mu^4 - 30\mu^2 + 3}{8},$$

$$P_5 = \frac{9.7}{2.4}\mu^5 - \frac{7.5}{2.2}\mu^3 + \frac{5.3}{2.4}\mu$$

$$= \frac{63\mu^5 - 70\mu^3 + 15\mu}{8},$$

$$P_6 = \frac{11.9.7}{2.4.6}\mu^6 - \frac{9.7.5}{2.4\times 2}\mu^4 + \frac{7.5.3}{2\times 2.4}\mu^2 - \frac{5.3.1}{2.4.6}$$

$$= \frac{231\mu^6 - 315\mu^4 + 105\mu^2 - 5}{16},$$

$$P_7 = \frac{13.11.9}{2.4.6}\mu^7 - \frac{11.9.7}{2.4\times 2}\mu^5 + \frac{9.7.5}{2\times 2.4}\mu^3 - \frac{7.5.3}{2.4.6}\mu$$

$$= \frac{429\mu^7 - 693\mu^5 + 315\mu^3 - 35\mu}{16},$$

$$P_8 = \frac{15.13.11.9}{2.4.6.8}\mu^8 - \frac{13.11.9.7}{2.4.6\times 2}\mu^6 + \frac{11.9.7.5}{2.4\times 2.4}\mu^4 - \frac{9.7.5.3}{2\times 2.4.6}\mu^2 + \frac{7.5.3.1}{2.4.6.8}$$

$$= \frac{6435\mu^8 - 12012\mu^6 + 6930\mu^4 - 1260\mu^2 + 35}{128},$$

$$P_9 = \frac{17.15.13.11}{2.4.6.8}\mu^9 - \frac{15.13.11.9}{2.4.6\times 2}\mu^7 + \frac{13.11.9.7}{2.4\times 2.4}\mu^5 - \frac{11.9.7.5}{2\times 2.4.6}\mu^3 + \frac{9.7.5.3}{2.4.6.8}\mu$$

$$= \frac{12155\mu^9 - 25740\mu^7 + 18018\mu^5 - 4620\mu^3 + 315\mu}{128},$$

$$P_{10} = \frac{19.17.15.13.11}{2.4.6.8.10}\mu^{10} - \frac{17.15.13.11.9}{2.4.6.8\times 2}\mu^8 + \frac{15.13.11.9.7}{2.4.6\times 2.4}\mu^6$$

$$-\frac{13.11.9.7.5}{2.4\times 2.4.6}\mu^4+\frac{11.9.7.5.3}{2\times 2.4.6.8}\mu^2-\frac{9.7.5.3.1}{2.4.6.8.10}$$

$$=\frac{46189\mu^{10}-109395\mu^8+90090\mu^6-30030\mu^4+3465\mu^2-63}{256}.$$

It will be observed that, when these fractions are reduced to their lowest terms, the denominators are in all cases powers of 2, the other factors being cancelled by corresponding factors in the numerator. The power of 2, in the denominator of P_i, is that which enters as a factor into the continued product $1.2\ldots i$.

16. We have seen that $\int_{-1}^{1}\mu^m P_i.d\mu=0$, if m be any integer less than i.

It will easily be seen that if $m+i$ be an odd number, the values of $\int\mu^m P_i.d\mu$ are the same, whether μ be put $=1$ or -1; but if $m+i$ be an even number, the values of $\int\mu^m P_i.d\mu$ corresponding to these limits are equal and opposite. Hence, ($m+i$ being even)

$$\int_{-1}^{1}\mu^m P_i.d\mu=2\int_0^1\mu^m P_i.d\mu,$$

and then $$\int_0^1\mu^m P_i.d\mu=0,\ \text{if}\ m=i-2,\ i-4\ldots\ldots$$

We may proceed to investigate the value of $\int_0^1\mu^m P_i.d\mu$, if m have any other value. For this purpose, resuming the notation of the equations of Art. 13, we see that, putting $\theta=m+1$, and $\omega=\infty$, we have

$$\frac{x_i}{a_i+m+1}+\frac{x_{i-2}}{a_{i-2}+m+1}+\ldots\ldots+\frac{x_{i-2s}}{a_{i-2s}+m+1}+\ldots\ldots$$

$$=\frac{(m+1-a_i)(m+1-a_{i-2})\ldots(m+1-a_{i-2s})\ldots}{(a_i+m+1)(a_{i-2}+m+1)\ldots(a_{i-2s}+m+1)\ldots};$$

and therefore, putting $x_i = A_i \ldots$, $a_i = i \ldots$, $\alpha_i = i - 1 \ldots$, we get

$$\int_0^1 \mu^m P_i . d\mu = \frac{A_i}{i+m+1} + \frac{A_{i-2}}{i-2+m+1} + \ldots + \frac{A_{i-2s}}{i-2s+m+1} + \ldots$$

$$= \frac{(m-i+2)(m-i+4)\ldots(m-1)}{(m+i+1)(m+i-1)\ldots(m+4)(m+2)} \text{ if } i \text{ be odd,}$$

and $$= \frac{(m-i+2)(m-i+4)\ldots m}{(m+i+1)(m+i-1)\ldots(m+3)(m+1)} \text{ if } i \text{ be even.}$$

In the particular case in which $m = i$, we get

$$\int_0^1 \mu^i P_i d\mu = \frac{2.4\ldots(i-1)}{(2i+1)(2i-1)\ldots(i+4)(i+2)} \; (i \text{ odd}),$$

and $$= \frac{2.4\ldots i}{(2i+1)(2i-1)\ldots(i+3)(i+1)} \; (i \text{ even}).$$

17. We may apply these formulæ to develope any positive integral power of μ in a series of zonal harmonics, as we proceed to shew.

Suppose that m is a positive integer, and that μ^m is developed in such a series, the coefficient of P_i being C_i, so that

$$\mu^m = \Sigma C_i P_i ;$$

then, multiplying both sides of this equation by P_i and integrating between the limits -1 and 1, all the terms on the right-hand side will disappear except $\int_{-1}^1 C_i P_i^2 d\mu$, which will become equal to $\frac{2}{2i+1} C_i$.

Hence $$C_i = \frac{2i+1}{2} \int_{-1}^1 \mu^m P_i d\mu,$$

which is equal to 0, if $m + i$ be odd. Hence no terms appear unless $m + i$ be even. In this case we have

$$C_i = \frac{2i+1}{2} \int_{-1}^1 \mu^m P_i d\mu$$

$$= (2i+1) \int_0^1 \mu^m P_i d\mu.$$

Hence the formula just investigated gives

$$C_i = (2i+1)\frac{(m-i+2)(m-i+4)\dots(m-1)}{(m+i+1)(m+i-1)\dots(m+4)(m+2)}$$

if i be odd, and

$$C_i = (2i+1)\frac{(m-i+2)(m-i+4)\dots m}{(m+i+1)(m+i-1)\dots(m+3)(m+1)}$$

if i be even.

Therefore if m be odd,

$$\mu^m = (2m+1)\frac{2.4.6\dots(m-1)}{(2m+1)(2m-1)\dots(m+4)(m+2)}P_m + \dots$$

$$+7\frac{m-1}{(m+4)(m+2)}P_3 + \frac{3}{m+2}P_1.$$

If m be even,

$$\mu^m = (2m+1)\frac{2.4.6\dots m}{(2m+1)(2m-1)\dots(m+3)(m+1)}P_m + \dots$$

$$+5\frac{m}{(m+3)(m+1)}P_2 + \frac{1}{m+1}P_0.$$

Hence, putting for m successively 0, 1, 2 ... 10, we get

$$\mu^0 = P_0,$$

$$\mu^1 = \frac{3}{3}P_1 = P_1,$$

$$\mu^2 = 5\frac{2}{5.3}P_2 + \frac{1}{3}P_0 = \frac{2}{3}P_2 + \frac{1}{3}P_0,$$

$$\mu^3 = 7\frac{2}{7.5}P_3 + \frac{3}{5}P_1$$

$$= \frac{2}{5}P_3 + \frac{3}{5}P_1,$$

$$\mu^4 = 9\frac{2.4}{9.7.5}P_4 + 5\frac{4}{7.5}P_2 + \frac{1}{5}P_0$$

$$= \frac{8}{35}P_4 + \frac{4}{7}P_2 + \frac{1}{5}P_0,$$

$$\mu^5 = 11\frac{2.4}{11.9.7}P_5 + 7\frac{4}{9.7}P_3 + \frac{3}{7}P_1$$

$$= \frac{8}{63}P_5 + \frac{4}{9}P_3 + \frac{3}{7}P_1,$$

$$\mu^6 = 13\frac{2.4.6}{13.11.9.7}P_6 + 9\frac{4.6}{11.9.7}P_4 + 5\frac{6}{9.7}P_2 + \frac{1}{7}P_0$$

$$= \frac{16}{231}P_6 + \frac{24}{77}P_4 + \frac{10}{21}P_2 + \frac{1}{7}P_0,$$

$$\mu^7 = 15\frac{2.4.6}{15.13.11.9}P_7 + 11\frac{4.6}{13.11.9}P_5 + 7\frac{6}{11.9}P_3 + \frac{3}{9}P_1,$$

$$= \frac{16}{429}P_7 + \frac{8}{39}P_5 + \frac{14}{33}P_3 + \frac{1}{3}P_1,$$

$$\mu^8 = 17\frac{2.4.6.8}{17.15.13.11.9}P_8 + 13\frac{4.6.8}{15.13.11.9}P_6$$

$$+ 9\frac{6.8}{13.11.9}P_4 + 5\frac{8}{11.9}P_2 + \frac{1}{9}P_0,$$

$$= \frac{128}{6435}P_8 + \frac{64}{495}P_6 + \frac{48}{143}P_4 + \frac{40}{99}P_2 + \frac{1}{9}P_0,$$

$$\mu^9 = 19\frac{2.4.6.8}{19.17.15.13.11}P_9 + 15\frac{4.6.8}{17.15.13.11}P_7$$

$$+ 11\frac{6.8}{15.13.11}P_5 + 7\frac{8}{13.11}P_3 + 3\frac{1}{11}P_1,$$

$$= \frac{128}{12155}P_9 + \frac{192}{2431}P_7 + \frac{16}{65}P_5 + \frac{56}{143}P_3 + \frac{3}{11}P_1,$$

$$\mu^{10} = 21\frac{2.4.6.8.10}{21.19.17.15.13.11}P_{10} + 17\frac{4.6.8.10}{19.17.15.13.11}P_8$$

$$+ 13\frac{6.8.10}{17.15.13.11}P_6 + 9\frac{8.10}{15.13.11}P_4 + 5\frac{10}{13.11}P_2 + \frac{1}{11}P_0$$

$$= \frac{256}{46189}P_{10} + \frac{128}{2717}P_8 + \frac{32}{187}P_6 + \frac{48}{143}P_4 + \frac{50}{143}P_2 + \frac{1}{11}P_0.$$

18. Any zonal harmonic P_i may be expressed in a finite series of cosines of multiples of θ, these multiples being $i\theta$, $(i-2)\theta$.... Thus

$$(1-2\mu h+h^2)^{-\frac{1}{2}} = P_0 + P_1 h + \ldots + P_i h^i + \ldots;$$

therefore, writing $\cos\theta$ for μ, and observing that

$$1-2\cos\theta h + h^2 = (1-h\epsilon^{\sqrt{-1}\theta})(1-h\epsilon^{-\sqrt{-1}\theta}),$$

we obtain

$$(1-h\epsilon^{\sqrt{-1}\theta})^{-\frac{1}{2}}(1-h\epsilon^{-\sqrt{-1}\theta})^{-\frac{1}{2}} = P_0 + P_1 h + \ldots + P_i h^i + \ldots$$

or

$$\left(1+\frac{1}{2}h\epsilon^{\sqrt{-1}\theta}+\frac{1.3}{2.4}h^2\epsilon^{\sqrt{-1}2\theta}+\ldots + \frac{1.3\ldots(2i-1)}{2.4\ldots 2i}h^i\epsilon^{\sqrt{-1}i\theta}+\ldots\right)$$

$$\times\left(1+\frac{1}{2}h\epsilon^{-\sqrt{-1}\theta}+\frac{1.3}{2.4}h^2\epsilon^{-\sqrt{-1}2\theta}+\ldots + \frac{1.3\ldots(2i-1)}{2.4\ldots 2i}h^i\epsilon^{-\sqrt{-1}i\theta}+\ldots\right)$$

$$= P_0 + P_1 h + \ldots + P_i h^i + \ldots$$

whence, equating coefficients of h^i,

$$P_i = \frac{1.3\ldots(2i-1)}{2.4\ldots 2i}2\cos i\theta + \frac{1.3\ldots(2i-3)}{2.4\ldots(2i-2)}\frac{1}{2}2\cos(i-2)\theta + \frac{1.3\ldots(2i-5)}{2.4\ldots(2i-4)}\frac{1.3}{2.4}2\cos(i-4)\theta + \ldots.$$

the last term being $\left\{\frac{1.3\ldots(i-1)}{2.4\ldots i}\right\}^2$ if i be even, and

$$\frac{1.3\ldots(i+1)}{2.4\ldots(i+1)}\,\frac{1.3\ldots(i-2)}{2.4\ldots(i-1)}\,2\cos\theta, \text{ if } i \text{ be odd.}$$

19. Let us next proceed to investigate the value of

$$\int_0^\pi P_i \cos m\theta \sin\theta\, d\theta.$$

This might be done, by direct integration, from the above expression. Or we may proceed as follows.

The above value of P_i when multiplied by $\cos m\theta \sin\theta$ (that is by $\frac{1}{2}\{\sin(m+1)\theta - \sin(m-1)\theta\}$) will consist of a series of sines of angles of the form $\{i - 2n \pm (m \pm 1)\}\theta$, that is of even or odd multiples of θ, as $i+m$ is odd or even. Therefore, when integrated between the limits 0 and π it will vanish, if $i+m$ be odd. We may therefore limit ourselves to the case in which $i+m$ is even.

Again, since $\cos m\theta$ can be expressed in a series of powers of $\cos\theta$, and the highest power involved in such an expression is $\cos^m\theta$, it follows that the highest zonal harmonic in the development of $\cos m\theta$ will be P_m. Hence $\int_0^\pi P_i \cos m\theta \sin\theta\, d\theta$ will be $=0$, if m be less than i.

Now, writing

$$P_i = C_i \cos i\theta + C_{i-2}\cos(i-2)\theta + \ldots$$

we see that $P_i \cos m\theta \sin\theta\, d\theta$ will consist of a series of sines of angles of the forms $(m+i+1)\theta$, $(m+i-1)\theta \ldots$ down to $(m-i-1)\theta$, there being no term involving $m\theta$, since the coefficient of such a term must be zero. Hence

$$\int_0^\pi P_i \cos m\theta \sin\theta\, d\theta,$$

will consist of a series of fractions whose denominators involve the factors $m+i+1, m+i-1 \ldots m-i-1$ respectively. Therefore when reduced to a common denominator, the result will involve in its denominator the factor

$$(m+i+1)(m+i-1)\ldots(m+1)(m-1)\ldots(m-i-1)$$

if m be even, and

$$(m+i+1)(m+i-1)\ldots(m+2)(m-2)\ldots(m-i-1)$$

if m be odd.

For the numerator we may observe that since

$$\int_0^\pi P_i \cos m\,\theta \sin\theta d\theta$$

vanishes if m be less than i, it must involve the factors $m-(i-2)$, $m-(i-4)\ldots m+(i-2)$, and that it does not change sign with m. Hence it will involve the factor

$$\{m-(i-2)\}\{m-(i-4)\}\ldots(m-2)\,m^2\,(m+2)\ldots(m+i-2)$$

if m be even, and

$$\{m-(i-2)\}\{m-(i-4)\}\ldots(m-1)(m+1)\ldots(m+i-2)$$

if m be odd.

To determine the factor independent of m, we may proceed as follows:

$$P_i = C_i \cos i\theta + C_{i-2}\cos(i-2)\,\theta + \ldots;$$

$$\therefore P_i \cos m\theta = \frac{1}{2}\,C_i\{\cos(m+i)\,\theta + \cos(m-i)\,\theta\}$$
$$+\frac{1}{2}\,C_{i-2}\{\cos(m+i-2)\,\theta+\cos(m-i+2)\,\theta\}+\ldots;$$

$$\therefore P_i\cos m\theta\sin\theta = \frac{1}{4}\,C_i\{\sin(m+i+1)\,\theta-\sin(m+i-1)\,\theta$$
$$+\sin(m-i+1)\,\theta-\sin(m-i-1)\,\theta\}$$
$$+\frac{1}{4}\,C_{i-2}\{\sin(m+i-1)\,\theta-\sin(m+i-3)\,\theta$$
$$+\sin(m-i+3)\,\theta-\sin(m-i+1)\,\theta\}+\ldots;$$

$$\therefore \int_0^\pi P_i\cos m\theta\sin\theta\,d\theta$$
$$=\frac{C_i}{2}\left\{\frac{1}{m+i+1}-\frac{1}{m+i-1}+\frac{1}{m-i+1}-\frac{1}{m-i-1}\right\}$$
$$+\frac{C_{i-2}}{2}\left\{\frac{1}{m+i-1}-\frac{1}{m+i-3}+\frac{1}{m-i+3}-\frac{1}{m-i+1}\right\}+\ldots$$
$$=C_i\left\{-\frac{i-1}{m^2-(i+1)^2}+\frac{i-1}{m^2-(i-1)^2}\right\}$$
$$+C_{i-2}\left\{-\frac{i-1}{m^2-(i-1)^2}+\frac{i-3}{m^2-(i-3)^2}\right\}+\ldots$$

Now, when m is very large as compared with i, this becomes

$$=-2\frac{C_i+C_{i-2}+\dots}{m^2}=-\frac{2}{m^2},$$

since $C_i+C_{i-2}+\dots=1$, as may be seen by putting $\theta=0$.

Hence $\int_0^\pi P_i \cos m\theta \sin\theta\, d\theta$ tends to the limit $-\frac{2}{m^2}$, as m is indefinitely increased.

The value of the factor involving m has been shewn above to be

$$\frac{\{m-(i-2)\}\{m-(i-4)\}\dots(m-2)\,m^2\,(m+2)\dots(m+i-2)}{\{m-(i+1)\}\{m-(i-1)\}\dots(m-1)(m+1)\dots(m+i+1)}$$

if m be even, and

$$\frac{\{m-(i-2)\}\{m-(i-4)\}\dots(m-1)(m+1)\dots(m+i-2)}{\{m-(i+1)\}\{m-(i-1)\}\dots(m-2)(m+2)\dots(m+i+1)}$$

if m be odd.

Each of these factors contains in its numerator two factors less than in its denominator. It approaches, therefore, when m is indefinitely increased, to the value $\frac{1}{m^2}$. Hence

$$\int_0^\pi P_i \cos m\theta \sin\theta\, d\theta$$

$$=-2\frac{\{m-(i-2)\}\{m-(i-4)\}\dots(m-2)m^2(m+2)\dots\{m+(i-2)\}}{\{m-(i+1)\}\{m-(i-1)\}\dots(m-1)(m+1)\dots\{m+(i+1)\}}$$

if m and i be even, and

$$=-2\frac{\{m-(i-2)\}\{m-(i-4)\}\dots(m-1)(m+1)\dots\{m+(i-2)\}}{\{m-(i+1)\}\{m-(i-1)\}\dots(m-2)(m+2)\dots\{m+(i+1)\}}$$

if m and i be odd.

In each of these expressions i may be any integer such that $m-i$ is even, i being *not greater* than m. Hence they will always be negative, *except when i is equal to m.*

20. We may apply these expressions to develop $\cos m\theta$ in a series of zonal harmonics.

Assume

$$\cos m\theta = B_m P_m + B_{m-2} P_{m-2} + \dots + B_i P_i + \dots$$

Multiply by $P_i \sin\theta$, and integrate between the limits 0 and π, and we get

$$-2\frac{\{m-(i-2)\}\{m-(i-4)\}\dots\{m+(i-2)\}}{\{m-(i+1)\}\{m-(i-1)\}\dots\{m+(i+1)\}} = \frac{2}{2i+1}B_i.$$

Hence

$$B_i = -(2i+1)\frac{\{m-(i-2)\}\{m-(i-4)\}\dots\{m+(i-2)\}}{\{m-(i+1)\}\{m-(i-1)\}\dots\{m+(i+1)\}}$$

Hence, putting m successively $= 0, 1, 2, \dots 10$,

$$\cos 0\theta = P_0;$$

$$\cos\theta = P_1;$$

$$\cos 2\theta = -5\frac{2^2}{-1.1.3.5}P_2 - \frac{1}{3}P_0$$

$$= \frac{4}{3}P_2 - \frac{1}{3}P_0;$$

$$\cos 3\theta = -7\frac{2.4}{-1.1.5.7}P_3 - 3\frac{1}{1.5}P_1$$

$$= \frac{8}{5}P_3 - \frac{3}{5}P_1;$$

$$\cos 4\theta = -9\frac{2.4^2.6}{-1.1.3.5.7.9}P_4 - 5\frac{4^2}{+1.3.5.7}P_2$$

$$-1\frac{1}{3.5}P_0$$

$$= \frac{64}{35}P_4 - \frac{16}{21}P_2 - \frac{1}{15}P_0;$$

$$\cos 5\theta = -11\frac{2.4.6.8}{-1.1.3.7.9.11}P_5 - 7\frac{4.6}{1.3.7.9}P_3$$
$$-3\frac{1}{3.7}P_1$$
$$= \frac{128}{63}P_5 - \frac{8}{9}P_3 - \frac{1}{7}P_1;$$

$$\cos 6\theta = -13\frac{2.4.6^2.8.10}{-1.1.3.5.7.9.11.13}P_6$$
$$-9\frac{4.6^2.8}{1.3.5.7.9.11}P_4 - 5\frac{6^2}{3.5.7.9}P_2 - \frac{1}{5.7}P_0$$
$$= \frac{512}{231}P_6 - \frac{384}{385}P_4 - \frac{4}{21}P_2 - \frac{1}{35}P_0;$$

$$\cos 7\theta = -15\frac{2.4.6.8.10.12}{-1.1.3.5.9.11.13.15}P_7$$
$$-11\frac{4.6.8.10}{1.3.5.9.11.13}P_5 - 7\frac{6.8}{3.5.9.11}P_3 - \frac{3}{5.9}P_1$$
$$= \frac{1024}{429}P_7 - \frac{128}{117}P_5 - \frac{112}{495}P_3 - \frac{1}{15}P_1;$$

$$\cos 8\theta = -17\frac{2.4.6.8^2.10.12.14}{-1.1.3.5.7.9.11.13.15.17}P_8$$
$$-13\frac{4.6.8^2.10.12}{1.3.5.7.9.11.13.15}P_6 - 9\frac{6.8^2.10}{3.5.7.9.11.13}P_4$$
$$-5\frac{8^2}{5.7.9.11}P_2 - \frac{1}{7.9}P_0$$
$$= \frac{16384}{6435}P_8 - \frac{4096}{3465}P_6 - \frac{256}{1001}P_4 - \frac{64}{693}P_2 - \frac{1}{63}P_0;$$

$$\cos 9\theta = -19\frac{2.4.6.8.10.12.14.16}{-1.1.3.5.7.11.13.15.17.19}P_9$$
$$-15\frac{4.6.8.10.12.14}{1.3.5.7.11.13.15.17}P_7 - 11\frac{6.8.10.12}{3.5.7.11.13.15}P_5$$
$$-7\frac{8.10}{5.7.11.13}P_3 - 3\frac{1}{7.11}P_1$$

$$=\frac{32768}{12155}P_9-\frac{3072}{2431}P_7-\frac{128}{455}P_5-\frac{16}{143}P_3-\frac{3}{77}P_1;$$

$$\cos 10\theta=-21\frac{2.4.6.8.10^2.12.14.16.18}{-1.1.3.5.7.9.11.13.15.17.19.21}P_{10}$$

$$-17\frac{4.6.8.10^2.12.14.16}{1.3.5.7.9.11.13.15.17.19}P_8$$

$$-13\frac{6.8.10^2.12.14}{3.5.7.9.11.13.15.17}P_6-9\frac{8.10^2.12}{5.7.9.11.13.15}P_4$$

$$-5\frac{10^2}{7.9.11.13}P_2-\frac{1}{9.11}P_0$$

$$=\frac{131072}{46189}P_{10}-\frac{32768}{24453}P_8-\frac{512}{1683}P_6-\frac{128}{1001}P_4-\frac{500}{9009}P_2$$

$$-\frac{1}{99}P_0.$$

21. The present will be a convenient opportunity for investigating the development of $\sin\theta$ in a series of zonal harmonics. Since $\sin\theta=(1-\mu^2)^{\frac{1}{2}}$, it will be seen that the series must be infinite, and that no zonal harmonic of an odd order can enter. Assume then

$$\sin\theta=C_0P_0+C_2P_2+\ldots+C_iP_i+\ldots$$

i being any even integer.

Multiplying by P_i, and integrating with respect to μ between the limits -1 and $+1$, we get

$$\int_{-1}^{1}P_i\sin\theta\,d\mu=\frac{2}{2i+1}C_i;$$

$$\therefore\ C_i=\frac{2i+1}{2}\int_{-1}^{1}P_i\sin\theta\,d\mu$$

$$=\frac{2i+1}{2}\int_0^{\pi}P_i\sin^2\theta\,d\theta,$$

supposing P_i expressed in terms of the cosines of θ and its multiples

$$=\frac{2i+1}{4}\int_0^{\pi}P_i(1-\cos 2\theta)\,d\theta.$$

Hence, putting $i=0$,

$$C_0=\frac{1}{4}\int_0^\pi (1-\cos 2\theta)\,d\theta=\frac{\pi}{4}.$$

Putting $i=2$, and observing that $P_2=\frac{1}{4}+\frac{3}{4}\cos 2\theta$,

$$C_2=\frac{5}{4}\int_0^\pi \frac{(1+3\cos 2\theta)(1-\cos 2\theta)}{4}\,d\theta$$

$$=\frac{5}{16}\int_0^\pi \left\{1+2\cos 2\theta-\frac{3}{2}(1+\cos 4\theta)\right\}d\theta$$

$$=-\frac{5}{32}\pi.$$

For values of i exceeding 2, we observe, that if we write for P_i the expression investigated in Art. 18, the only part of the expression $\int_0^\pi P_i(1-\cos 2\theta)\,d\theta$ which does not vanish will arise either from the terms in P_i which involve $\cos 2\theta$, or from those which are independent of θ. We have therefore

$$C_i=\frac{2i+1}{4}\int_0^\pi\left[\frac{1.3\ldots(i+1)}{2.4\ldots(i+2)}\frac{1.3\ldots(i-3)}{2.4\ldots(i-2)}2\cos 2\theta+\left\{\frac{1.3\ldots(i-1)}{2.4\ldots\quad i}\right\}^2\right](1-\cos 2\theta)\,d\theta$$

$$=\frac{2i+1}{4}\cdot\frac{1.3\ldots(i-1)}{2.4\ldots\quad i}\,\frac{1.3\ldots(i-3)}{2.4\ldots(i-2)}\int_0^\pi\left(\frac{i-1}{i}+\frac{i+1}{i+2}2\cos 2\theta\right)(1-\cos 2\theta)\,d\theta$$

$$=\frac{2i+1}{4}\,\frac{1.3\ldots(i-1)}{2.4\ldots\quad i}\,\frac{1.3\ldots(i-3)}{2.4\ldots(i-2)}\,\pi\left(\frac{i-1}{i}-\frac{i+1}{i+2}\right)$$

$$=-\pi\frac{2i+1}{2}\,\frac{1.3\ldots(i-1)}{2.4\ldots i(i+2)}\,\frac{1.3\ldots(i-3)}{2.4\ldots(i-2)\,i}.$$

Hence $\sin\theta=\frac{\pi}{4}P_0-\frac{5\pi}{32}P_2-\ldots$

$$-\frac{(2i+1)\pi}{2}\,\frac{1.3\ldots(i-1)}{2.4\ldots i(i+2)}\,\frac{1.3\ldots(i-3)}{2.4\ldots(i-2)\,i}P_i-\ldots$$

i being any even integer.

22. It will be seen that $\frac{dP_i}{d\mu}$, being a rational and integral function of μ^{i-1}, μ^{i-3}..., must be expressible in terms of P_{i-1}, P_{i-3}... To determine this expression, assume

$$\frac{dP_i}{d\mu} = C_{i-1} P_{i-1} + C_{i-3} P_{i-3} + \dots + C_m P_m + \dots$$

then multiplying by P_m, and integrating with respect to μ from -1 to $+1$,

$$\int_{-1}^{1} P_m \frac{dP_i}{d\mu} d\mu = C_m \int_{-1}^{1} P_m^2 d\mu = \frac{2}{2m+1} C_m.$$

And $$\int P_m \frac{dP_i}{d\mu} d\mu = P_m P_i - \int P_i \frac{dP_m}{d\mu} d\mu.$$

Now, since $i > m$,

$$\int_{-1}^{1} P_i \frac{dP_m}{d\mu} d\mu = 0;$$

$$\therefore \int_{-1}^{1} P_m \frac{dP_i}{d\mu} d\mu = [P_m P_i]^1 - [P_m P_i]^{-1} = 2,$$

since either m or i must be odd, and therefore either P_m or $P_i = -1$, when $\mu = -1$;

$$\therefore 2 = \frac{2}{2m+1} C_m, \text{ or } C_m = 2m + 1;$$

$$\therefore \frac{dP_i}{d\mu} = (2i-1) P_{i-1} + (2i-5) P_{i-3} + (2i-9) P_{i-5} + \dots$$

Hence $$\frac{dP_i}{d\mu} - \frac{dP_{i-2}}{d\mu} = (2i-1) P_{i-1}.$$

23. From this equation we deduce

$$P_i - P_{i-2} = -(2i-1) \int_{\mu}^{1} P_{i-1} d\mu,$$

the limits μ and 1 being taken, in order that $P_i - P_{i-2}$ may be equal to 0 at the superior limit.

Now, recurring to the fundamental equation for a zonal harmonic, we see that

$$\int_\mu^1 P_{i-1}\,d\mu = \frac{1}{i(i-1)}(1-\mu^2)\frac{dP_{i-1}}{d\mu};$$

$$\therefore\; P_i - P_{i-2} = -\frac{2i-1}{i(i-1)}(1-\mu^2)\frac{dP_{i-1}}{d\mu}$$

$$= -\left(\frac{1}{i}+\frac{1}{i-1}\right)(1-\mu^2)\frac{dP_{i-1}}{d\mu}.$$

24. We have already seen that $\int_{-1}^1 P_i P_m\,d\mu = 0$, i and m being different positive integers. Suppose now that it is required to find the value of $\int_\mu^1 P_i P_m\,d\mu$.

We have already seen (Art. 10) that

$$\int_\mu^1 P_i P_m\,d\mu = \frac{(1-\mu^2)\left(P_m\dfrac{dP_i}{d\mu} - P_i\dfrac{dP_m}{d\mu}\right)}{(i-m)(i+m+1)}.$$

And, from above,

$$(1-\mu^2)\frac{dP_i}{d\mu} = -\frac{i(i+1)}{2i+1}(P_{i+1}-P_{i-1});$$

$$(1-\mu^2)\frac{dP_m}{d\mu} = -\frac{m(m+1)}{2m+1}(P_{m+1}-P_{m-1}).$$

$$\therefore \int_\mu^1 P_i P_m\,d\mu = \frac{1}{(i-m)(i+m+1)}\left\{\frac{m(m+1)}{2m+1}P_i(P_{m+1}-P_{m-1}) - \frac{i(i+1)}{2i+1}P_m(P_{i+1}-P_{i-1})\right\}.$$

25. We will next proceed to give two modes of expressing Zonal Harmonics, by means of Definite Integrals. The two expressions are as follows:

$$P_i = \frac{1}{\pi}\int_0^\pi \frac{d\vartheta}{\{\mu \pm (\mu^2-1)^{\frac{1}{2}}\cos\vartheta\}^{i+1}},$$

$$P_i = \frac{1}{\pi}\int_0^\pi \{\mu \pm (\mu^2-1)^{\frac{1}{2}}\cos\psi\}^i\, d\psi.$$

These we proceed to establish.

Consider the equation

$$\frac{1}{\pi}\int_0^\pi \frac{d\vartheta}{a - b\cos\vartheta} = \frac{1}{(a^2-b^2)^{\frac{1}{2}}}.$$

The only limitation upon the quantities denoted by a and b in this equation is that b^2 should not be greater than a^2. For, if b^2 be not greater than a^2, $\cos\vartheta$ cannot become equal to $\frac{a}{b}$ while ϑ increases from 0 to π, and therefore the expression under the integral sign cannot become infinite.

Supposing then that we write z for a, and $\sqrt{-1}\,\rho$ for b, we get

$$\frac{1}{\pi}\int_0^\pi \frac{d\vartheta}{z - \sqrt{-1}\,\rho\cos\vartheta} = \frac{1}{(z^2+\rho^2)^{\frac{1}{2}}}.$$

We may remark, in passing, that

$$\int_0^\pi \frac{d\vartheta}{z - \sqrt{-1}\,\rho\cos\vartheta} = \int_0^\pi \frac{d\vartheta}{z + \sqrt{-1}\,\rho\cos\vartheta}$$

$$= \int_0^\pi \frac{z\,d\vartheta}{z^2 + \rho^2\cos^2\vartheta},$$

and is therefore wholly real.

Supposing that $\rho^2 = x^2 + y^2$, and that $x^2 + y^2 + z^2 = r^2$, we thus obtain

$$\frac{1}{\pi}\int_0^\pi \frac{d\vartheta}{z - \sqrt{-1}\,\rho\cos\vartheta} = \frac{1}{r}.$$

Differentiate i times with respect to z, and there results

$$\frac{1}{\pi}\frac{d^i}{dz^i}\int_0^\pi \frac{d\vartheta}{z-\sqrt{-1}\,\rho\cos\vartheta} = \frac{d^i}{dz^i}\frac{1}{r} = \frac{P_i}{r^{i+1}}(-1)^i 1.2.3\ldots i.$$

Hence $P_i = \dfrac{r^{i+1}}{\pi}\,\dfrac{(-1)^i}{1.2.3\ldots i}\,\dfrac{d^i}{dz^i}\displaystyle\int_0^\pi \frac{d\vartheta}{z-\sqrt{-1}\,\rho\cos\vartheta}$

$$= \frac{r^{i+1}}{\pi}\int_0^\pi \frac{d\vartheta}{(z-\sqrt{-1}\,\rho\cos\vartheta)^{i+1}}.$$

In this, write μr for z, and $(1-\mu^2)^{\frac{1}{2}} r$ for ρ, and we get

$$P_i = \frac{1}{\pi}\int_0^\pi \frac{d\vartheta}{\{\mu-(\mu^2-1)^{\frac{1}{2}}\cos\vartheta\}^{i+1}},$$

which, writing $\pi-\vartheta$ for ϑ, gives

$$P_i = \frac{1}{\pi}\int_0^\pi \frac{d\vartheta}{\{\mu+(\mu^2-1)^{\frac{1}{2}}\cos\vartheta\}^{i+1}}.$$

26. Again, we have

$$\frac{1}{(a^2-b^2)^{\frac{1}{2}}} = \frac{1}{\pi}\int_0^\pi \frac{d\psi}{a-b\cos\psi}.$$

In this write $1-\mu h$ for a, and $\pm(\mu^2-1)^{\frac{1}{2}} h$ for b, which is admissible for all values of h from 0 up to $\mu-(\mu^2-1)^{\frac{1}{2}}$, and we obtain, since a^2-b^2 becomes $1-2\mu h+h^2$,

$$\frac{1}{(1-2\mu h+h^2)^{\frac{1}{2}}} = \frac{1}{\pi}\int_0^\pi \frac{d\psi}{1-\mu h\mp(\mu^2-1)^{\frac{1}{2}} h\cos\psi}$$

$$= \frac{1}{\pi}\int_0^\pi \frac{d\psi}{1-\{\mu\pm(\mu^2-1)^{\frac{1}{2}}\cos\psi\} h};$$

$$\therefore 1+P_1 h+\ldots+P_i h^i+\ldots$$

$$= \frac{1}{\pi}\int_0^\pi d\psi\,[1+\{\mu\pm(\mu^2-1)^{\frac{1}{2}}\cos\psi\} h+\ldots$$

$$+\{\mu\pm(\mu^2-1)^{\frac{1}{2}}\cos\psi\}^i h^i+\ldots].$$

Hence, equating coefficients of h^i,

$$P_i = \frac{1}{\pi}\int_0^\pi \{\mu \pm (\mu^2-1)^{\frac{1}{2}} \cos\psi\}^i \, d\psi.$$

The equality of the two expressions thus obtained for P_i is in harmony with the fact to which attention has already been directed, that the value of P_i is unaltered if $-(i+1)$ be written for i.

27. The equality of the two definite integrals which thus present themselves may be illustrated by the following geometrical considerations.

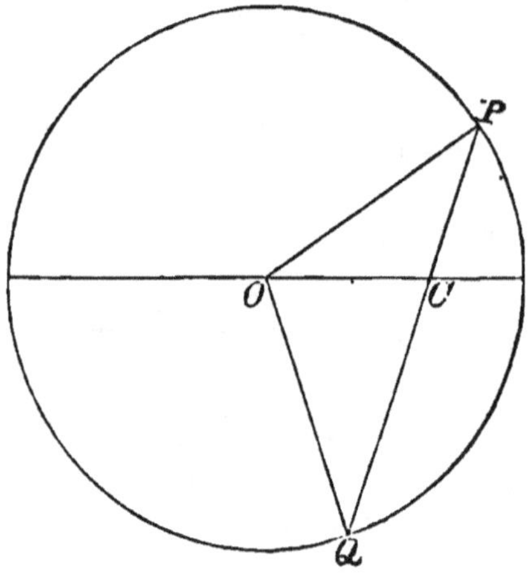

Let O be the centre of a circle, radius a, C any point within the circle, PCQ any chord drawn through C, and let $OC=b$, $COP=\vartheta$, $COQ=\psi$. Then $CP^2=a^2+b^2-2ab\cos\vartheta$, $CQ^2=a^2+b^2-2ab\cos\psi$. Hence

$$(a^2+b^2-2ab\cos\vartheta)(a^2+b^2-2ab\cos\psi)=(a^2-b^2)^2;$$

$$\therefore \frac{\sin\vartheta\, d\vartheta}{a^2+b^2-2ab\cos\vartheta}+\frac{\sin\psi\, d\psi}{a^2+b^2-2ab\cos\psi}=0.$$

Again, since the angles OPC, OQC are equal to one another,

$$\frac{\sin\theta}{CP}=\frac{\sin OPC}{OC}=\frac{\sin OQC}{OC}=\frac{\sin\psi}{CQ};$$

$$\therefore \frac{\sin\vartheta}{(a^2+b^2-2ab\cos\vartheta)^{\frac{1}{2}}}=\frac{\sin\psi}{(a^2+b^2-2ab\cos\psi)^{\frac{1}{2}}},$$

whence $$\frac{d\vartheta}{(a^2+b^2-2ab\cos\vartheta)^{\frac{1}{2}}}+\frac{d\psi}{(a^2+b^2-2ab\cos\psi)^{\frac{1}{2}}}=0.$$

And $$\frac{(a^2-b^2)^{2i+1}}{(a^2+b^2-2ab\cos\vartheta)^{i+\frac{1}{2}}}=(a^2+b^2-2ab\cos\psi)^{i+\frac{1}{2}};$$

$$\therefore (a^2-b^2)^{2i+1}\frac{d\vartheta}{(a^2+b^2-2ab\cos\vartheta)^{i+1}}=-(a^2+b^2-2ab\cos\psi)^i\,d\psi.$$

In this, write $a^2+b^2=\mu$, $2ab=\mp(\mu^2-1)^{\frac{1}{2}}$, which gives $a^2-b^2=1$, and we get

$$\frac{d\vartheta}{\{\mu\pm(\mu^2-1)^{\frac{1}{2}}\cos\vartheta\}^{i+1}}=-\{\mu\pm(\mu^2-1)^{\frac{1}{2}}\cos\psi\}^i\,d\psi.$$

We also see, by reference to the figure, that as ϑ increases from 0 to π, ψ diminishes from π to 0. Hence

$$\int_0^\pi\frac{d\vartheta}{\{\mu\pm(\mu^2-1)^{\frac{1}{2}}\cos\vartheta\}^{i+1}}=\int_0^\pi\{\mu\pm(\mu^2-1)^{\frac{1}{2}}\cos\psi\}^i\,d\psi.$$

28. From the last definite integral, we may obtain an expansion of P_i in terms of $\cos\theta$ and $\sin\theta$. Putting $\mu=\cos\theta$, we get

$$P_i=\frac{1}{2\pi}\int_0^\pi[\{\cos\theta+\sqrt{-1}\cos\psi\sin\theta)\}^i$$

$$+\{\cos\theta-\sqrt{-1}\cos\psi\sin\theta\}^i]\,d\psi$$

$$=\frac{1}{\pi}\int_0^\pi\{(\cos\theta)^i-\frac{i(i-1)}{1.2}\cos^2\psi(\cos\theta)^{i-2}(\sin\theta)^2+\dots$$

$$+(-1)^m \frac{i(i-1)\dots(i-2m+1)}{1.2\dots 2m} (\cos\psi^2)^m (\cos\theta)^{i-2m} (\sin\theta)^{2m}$$

$$+\dots\} \, d\psi.$$

Now $\int_0^\pi (\cos\psi)^{2m} d\psi = \pi \frac{(2m-1)(2m-3)\dots 1}{2m(2m-2)\dots 2}.$

And $\frac{i(i-1)\dots(i-2m+1)}{1.2\dots 2m} \frac{(2m-1)(2m-3)\dots 1}{2m(2m-2)\dots 2}$

$$= \frac{i(i-1)\dots(i-2m+1)}{(2.4\dots 2m)^2};$$

$$\therefore P_i = (\cos\theta)^i - \frac{i(i-1)}{2^2} (\cos\theta)^{i-2} (\sin\theta)^2 + \dots$$

$$+(-1)^m \frac{i(i-1)\dots(i-2m+1)}{(2.4\dots 2m)^2} (\cos\theta)^{i-2m} (\sin\theta)^{2m} + \dots$$

CHAPTER III.

APPLICATION OF ZONAL HARMONICS TO THE THEORY OF ATTRACTION. REPRESENTATION OF DISCONTINUOUS FUNCTIONS BY SERIES OF ZONAL HARMONICS.

1. WE shall, in this chapter, give some applications of Zonal Harmonics to the determination of the potential of a solid of revolution, symmetrical about an axis. When the value of this potential, at every point of the axis, is known, we can obtain, by means of these functions, an expression for the potential at any point which can be reached from the axis without passing through the attracting mass.

The simplest case of this kind is that in which the attracting mass is an uniform circular wire, of indefinitely small transverse section.

Let c be the radius of such a wire, ρ its density, k its transverse section. Then its mass, M, will be equal to $2\pi\rho ck$, and if its centre be taken as the origin, its potential at any point of its axis, distant z from its centre, will be $\dfrac{M}{(c^2+z^2)^{\frac{1}{2}}}$.

Now, this expression may be developed into either of the following series:

$$\frac{M}{c}\left\{1-\frac{1}{2}\frac{z^2}{c^2}+\frac{1.3}{2.4}\frac{z^4}{c^4}-\ldots+(-1)^i\frac{1.3\ldots(2i-1)}{2.4\ldots 2i}\frac{z^{2i}}{c^{2i}}+\ldots\right\}\ldots(1),$$

$$\frac{M}{z}\left\{1-\frac{1}{2}\frac{c^2}{z^2}+\frac{1.3}{2.4}\frac{c^4}{z^4}-\ldots+(-1)^i\frac{1.3\ldots(2i-1)}{2.4\ldots 2i}\frac{c^{2i}}{z^{2i}}+\ldots\right\}\ldots(2).$$

We must employ the series (1) if z be less than c, or if the attracted point lie within the sphere of which the ring is a great circle, and the series (2) if z be greater than c, or if the attracted point lie without this sphere.

Now, take any point whose distance from the centre is r, and let the inclination of this distance to the axis of the ring be θ. In accordance with the notation already employed, let $\cos\theta = \mu$. Then, the potential at this point will be given by one of the following series:

$$\frac{M}{c}\left\{P_0 - \frac{1}{2}P_2\frac{r^2}{c^2} + \frac{1.3}{2.4}P_4\frac{r^4}{c^4} - \dots \right.$$
$$\left. + (-1)^i \frac{1.3.5\dots(2i-1)}{2.4.6\dots 2i} P_{2i}\frac{r^{2i}}{c^{2i}} + \dots\right\} \dots (1'),$$

$$\frac{M}{r}\left\{P_0 - \frac{1}{2}P_2\frac{c^2}{r^2} + \frac{1.3}{2.4}P_4\frac{c^4}{r^4} - \dots \right.$$
$$\left. + (-1)^i \frac{1.3.5\dots(2i-1)}{2.4.6\dots 2i} P_{2i}\frac{c^{2i}}{r^{2i}} + \dots\right\} \dots (2').$$

For each of these expressions, when substituted for V, satisfies the equation $\nabla^2 V = 0$, and they become respectively equal to (1) and (2) when θ is put $= 0$, and consequently $r = z$. The expression (2′) also vanishes when r is infinitely great, and must therefore be employed for values of r greater than c, while (1′) becomes equal to (2′) when $r = c$, and will therefore denote the required potential for all values of r less than c.

These expressions may be reduced to other forms by means of the expressions investigated in Chap. 2, Art. 25, viz.

$$P_i = \frac{1}{\pi}\int_0^\pi (\mu + \sqrt{\mu^2 - 1}\cos\vartheta)^i \, d\vartheta,$$

$$\text{or } P_i = \frac{1}{\pi}\int_0^\pi (\mu + \sqrt{\mu^2 - 1}\cos\psi)^{-(i+1)} \, d\psi.$$

Substitute the first of these in (1′) and (observing that $\mu r = z$) we see that it assumes the form

$$\frac{M}{\pi c}\int_0^\pi \left\{1 - \frac{1}{2}\frac{\{z + (z^2 - r^2)^{\frac{1}{2}}\cos\vartheta\}^2}{c^2}\right.$$
$$\left. + \frac{1.3}{2.4}\frac{\{z + (z^2 - r^2)^{\frac{1}{2}}\cos\vartheta\}^4}{c^4} - \dots\right\} d\vartheta,$$

which is equivalent to

$$\frac{M}{\pi}\int_0^\pi \frac{d\vartheta}{[c^2+\{z+(z^2-r^2)^{\frac{1}{2}}\cos\vartheta\}^2]^{\frac{1}{2}}}.$$

The substitution of the last form of P_i in the series (2′) brings it into the form

$$\frac{M}{\pi}\int_0^\pi \left\{\frac{1}{z+(z^2-r^2)^{\frac{1}{2}}\cos\vartheta}-\frac{1}{2}\frac{c^2}{\{z+(z^2-r^2)^{\frac{1}{2}}\cos\vartheta\}^3}\right.$$
$$\left.+\frac{1.3}{2.4}\frac{c^4}{\{z+(z^2-r^2)^{\frac{1}{2}}\cos\vartheta\}^5}-\ldots\right\}d\vartheta,$$

which is equivalent to

$$\frac{M}{\pi}\int_0^\pi \frac{d\vartheta}{[\{z+(z^2-r^2)^{\frac{1}{2}}\cos\vartheta\}^2+c^2]^{\frac{1}{2}}}.$$

2. Suppose next that the attracting mass is a hollow shell of uniform density, whose exterior and interior bounding surfaces are both surfaces of revolution, their common axis being the axis of z. Let the origin be taken within the interior bounding surface; and suppose the potential, at any point of the axis within this surface, to be

$$A_0+A_1z+A_2z^2+\ldots+A_iz^i+\ldots$$

Then the potential at any point lying within the inner bounding surface will be

$$A_0P_0+A_1P_1r+A_2P_2r^2+\ldots+A_iP_ir^i+\ldots$$

For this expression, when substituted for V, satisfies the equation $\nabla^2V=0$; it also agrees with the given value of the potential for every point of the axis, lying within the inner bounding surface, and does not become infinite at any point within that surface.

Again, suppose the potential at any point of the axis without the outer bounding surface to be

$$\frac{B_0}{z}+\frac{B_1}{z^2}+\frac{B_2}{z^3}+\ldots+\frac{B_i}{z^{i+1}}+\ldots$$

Then the potential at any point lying without the outer bounding surface will be

$$\frac{B_0P_0}{r}+\frac{B_1P_1}{r^2}+\frac{B_2P_2}{r^3}+\ldots+\frac{B_iP_i}{r^{i+1}}+\ldots.$$

For this expression, when substituted for V, satisfies the equation $\nabla^2 V=0$; it also agrees with the given value of the potential for every point of the axis, lying without the outer bounding surface, and it does not become infinite at any point within that surface.

By the introduction of the expressions for zonal harmonics in the form of definite integrals, it will be found that if the value of either of these potentials for any point in the axis be denoted by $\phi(z)$, the corresponding value for any other point, which can be reached without passing through any portion of the attracting mass, will be

$$\frac{1}{\pi}\int_0^\pi \phi\{z+(z^2-r^2)^{\frac{1}{2}}\cos\vartheta\}\,d\vartheta.$$

3. We may next shew how to obtain, in terms of a series of zonal harmonics, an expression for the solid angle subtended by a circle at any point. We must first prove the following theorem.

The solid angle, subtended by a closed plane curve at any point, is proportional to the component attraction perpendicular to the plane of the curve, exercised upon the point by a lamina, of uniform density and thickness, bounded by the closed plane curve.

For, if dS be any element of such a lamina, r its distance from the attracted point, θ the inclination of r to the line perpendicular to the plane of the lamina, the elementary solid angle subtended by dS at the point will be

$$\frac{dS\cos\theta}{r^2}.$$

And the component attraction of the element of the lamina corresponding to dS in the direction perpendicular to its plane will be

$$\rho k\frac{dS}{r^2}\cos\theta,$$

ρ being the density of the lamina, k its thickness. Hence, for this element, the component attraction is to the solid angle as ρk to 1, and the same relation holding for every element of the lamina, we see that the component attraction of the whole lamina is to the solid angle subtended by the whole curve as ρk to 1.

Now, if the plane of xy be taken parallel to the plane of the lamina, and V be the potential of the lamina, its component attraction perpendicular to its plane will be $-\frac{dV}{dz}$. Now since V is a potential we have $\nabla^2 V=0$, whence $\frac{d}{dz}\nabla^2 V=0$, or $\nabla^2\left(\frac{dV}{dz}\right)=0$. Hence $\frac{dV}{dz}$ is itself a potential, and satisfies all the analytical conditions to which a potential is subject. It follows that, if the solid angle subtended by a closed plane curve at any point (x, y, z) be denoted by ω, ω will be a function of x, y, z, satisfying the equation $\nabla^2\omega = 0$. Hence, if the closed plane curve be a circle it follows that the magnitude of the solid angle which it subtends at any point may be obtained by first determining the solid angle which it subtends at any point of a line drawn through its centre perpendicular to its plane, and then deducing the general expression by the employment of zonal harmonics.

Now let O be the centre of the circle, Q any point on the line drawn through O perpendicular to the plane of the circle, E any point in the circumference of the circle. With centre Q, and radius QO, describe a circle, cutting QE in L. From L draw LN, perpendicular to QO.

Let $OE = c$, $OQ = z$.

$$\text{Then } EL = (c^2+z^2)^{\frac{1}{2}} - z,\quad ON = \frac{z}{(c^2+z^2)^{\frac{1}{2}}}\{(c^2+z^2)^{\frac{1}{2}} - z\}$$

$$= z - \frac{z^2}{(c^2+z^2)^{\frac{1}{2}}}.$$

And the solid angle subtended by the circle at Q

$$= 4\pi \frac{ON}{2z}$$

$$= 2\pi \left\{1 - \frac{z}{(c^2 + z^2)^{\frac{1}{2}}}\right\}.$$

To obtain the general expression for the solid angle subtended at any point, distant r from the centre, we first develope this expression in a converging series, proceeding by powers of z. This will be

$$2\pi \left\{1 - \frac{z}{c} + \frac{1}{2}\frac{z^3}{c^3} - \frac{1.3}{2.4}\frac{z^5}{c^5} + \ldots - (-1)^i \frac{1.3\ldots(2i-1)}{2.4\ldots 2i}\frac{z^{2i+1}}{c^{2i+1}} + \ldots\right\}$$

if z be less than c, and

$$2\pi \left\{\frac{1}{2}\frac{c^2}{z^2} - \frac{1.3}{2.4}\frac{c^4}{z^4} + \ldots - (-1)^i \frac{1.3\ldots(2i-1)}{2.4\ldots 2i}\frac{c^{2i}}{z^{2i}} + \ldots\right\}$$

if z be greater than c.

Hence, by similar reasoning to that already employed, we get, for the solid angle subtended at a point distant r from the centre,

$$2\pi \left\{P_0 - \frac{P_1 r}{c} + \frac{1}{2}\frac{P_3 r^3}{c^3} - \frac{1.3}{2.4}\frac{P_5 z^5}{c^5} + \ldots \right.$$

$$\left. - (-1)^i \frac{1.3\ldots(2i-1)}{2.4\ldots 2i}\frac{P_{2i+1} r^{2i+1}}{c^{2i+1}} + \ldots\right\}$$

if r be less than c, and

$$2\pi \left\{\frac{1}{2}\frac{P_1 c^2}{r^2} - \frac{1.3}{2.4}\frac{P_3 c^4}{r^4} + \ldots - (-1)^i \frac{1.3\ldots(2i-1)}{2.4\ldots 2i}\frac{P_{2i-1} c^{2i}}{r^{2i}} + \ldots\right\}$$

if r be greater than c.

4. We may deduce from this, expressions for the potential of a circular lamina, of uniform thickness and density, at an external point. For we see that, if V be the potential of such a lamina, k its thickness, and ρ its density, we have for a point on the axis,

$$-\frac{dV}{dz} = \rho k \,.\, 2\pi \left\{1 - \frac{z}{(c^2 + z^2)^{\frac{1}{2}}}\right\},$$

whence
$$V = 2\pi\rho k\{(c^2+z^2)^{\frac{1}{2}} - z\}$$
$$= \frac{M}{c^2}\{(c^2+z^2)^{\frac{1}{2}} - z\},$$

if M be the mass of the lamina.

Expanding this in a converging series, we get

$$V = \frac{M}{c^2}\left\{c - z + \frac{1}{2}\frac{z^2}{c} - \frac{1.1}{2.4}\frac{z^4}{c^3} + \frac{1.1.3}{2.4.6}\frac{z^6}{c^5} - \dots \right.$$
$$\left. -(-1)^i\frac{1.1.3\dots(2i-3)}{2.4.6\dots2i}\frac{z^{2i}}{c^{2i-1}} + \dots\right\}$$

if z be less than c, and

$$V = \frac{M}{c^2}\left\{\frac{1}{2}\frac{c^2}{z} - \frac{1.1}{2.4}\frac{c^4}{z^3} + \frac{1.1.3}{2.4.6}\frac{c^6}{z^5} - \dots \right.$$
$$\left. -(-1)^i\frac{1.1.3\dots(2i-3)}{2.4.6\dots2i}\frac{c^{2i}}{z^{2i-1}} + \dots\right\}$$

if z be greater than c.

Hence we obtain the following expressions for the potential of an uniform circular lamina at a point distant r from the centre of the lamina:

$$V = \frac{M}{c^2}\left\{P_0 c - P_1 r + \frac{1}{2}\frac{P_2 r^2}{c} - \frac{1.1}{2.4}\frac{P_4 r^4}{c^3} + \dots \right.$$
$$\left. -(-1)^i\frac{1.1.3\dots(2i-3)}{2.4.6\dots2i}\frac{P_{2i}r^{2i}}{c^{2i-1}} + \dots\right\}$$

if r be less than c, and

$$V = \frac{M}{c^2}\left\{\frac{1}{2}\frac{P_0 c^2}{r} - \frac{1.1}{2.4}\frac{P_2 c^4}{r^3} + \frac{1.1.3}{2.4.6}\frac{P_4 c^6}{r^5} - \dots \right.$$
$$\left. -(-1)^i\frac{1.1.3\dots(2i-3)}{2.4.6\dots2i}\frac{P_{2i-2}c^{2i}}{r^{2i-1}} + \dots\right\}$$

if r be greater than c.

It may be shewn that the solid angle may be expressed in the form

$$2\pi - 2\int_0^\pi \frac{z+(z^2-r^2)^{\frac{1}{2}}\cos\theta}{[c^2+\{z+(z^2-r^2)^{\frac{1}{2}}\cos\theta\}^2]^{\frac{1}{2}}}\,d\theta,$$

and the potential of the lamina in the form

$$\frac{M}{c^2}\frac{2}{\pi}\int_0^\pi [c^2+\{z+(z^2-r^2)^{\frac{1}{2}}\cos\theta\}^2]^{\frac{1}{2}}\,d\theta - \frac{Mr}{c^2}.$$

5. As another example, let it be required to determine the potential of a solid sphere, whose density varies inversely as the fifth power of the distance from a given external point O at any point of its mass.

It is proved by the method of inversion (see Thomson and Tait's *Natural Philosophy*, Vol. 1, Art. 518) that the potential at any external point P' will be equal to $\frac{M}{O'P'}$, O' being the image of O in the surface of the sphere, and M the mass of the sphere. We shall avail ourselves of this result to determine the potential at a given internal point.

Let C be the centre of the sphere, O the given external point. Join CO, and let it cut the surface of the sphere in A, and in CA take a point O', such that $CO\,.\,CO' = CA^2$. Then O' is the image of O.

Let P be any point in the body of the sphere, then we wish to find the potential of the sphere at P.

Take O as pole, and OC as prime radius, let $OP = r$, $POC = \theta$. Also let $CA = a$, $CO = c$.

Let the density of the sphere at its centre be ρ, then its density at P will be $\rho\frac{c^5}{r^5}$. Hence

$$M = 2\pi\iint \rho\frac{c^5}{r^5}r^2\sin\theta\,dr\,d\theta,$$

the limits of r being the two values of r which satisfy the equation of the surface of the sphere, viz.

$$r^2 + c^2 - 2cr\cos\theta = a^2,$$

and those of θ being 0 and $\sin^{-1}\frac{a}{c}$.

Hence, if r_1, r_2 be the two limiting values of r, we have

$$M = \frac{2\pi\rho c^5}{2}\int_0^{\sin^{-1}\frac{a}{c}}\left(\frac{1}{r_2^{\,2}} - \frac{1}{r_1^{\,2}}\right)\sin\theta\,d\theta.$$

Now $$\frac{1}{r_2^{\,2}} - \frac{1}{r_1^{\,2}} = \frac{2c\cos\theta}{c^2 - a^2}\left(\frac{1}{r_2} - \frac{1}{r_1}\right).$$

Also $$\frac{1}{r_2} + \frac{1}{r_1} = \frac{2c\cos\theta}{c^2 - a^2},$$

$$\frac{1}{r_1 r_2} = \frac{1}{c^2 - a^2};$$

$$\therefore \left(\frac{1}{r_2} - \frac{1}{r_1}\right) = 2\,\frac{\{c^2\cos^2\theta - (c^2 - a^2)\}^{\frac{1}{2}}}{c^2 - a^2}$$

$$= 2\,\frac{(a^2 - c^2\sin^2\theta)^{\frac{1}{2}}}{c^2 - a^2};$$

$$\therefore M = \frac{2\pi\rho c^5}{2}\,\frac{2c}{c^2 - a^2}\cdot\frac{2}{c^2 - a^2}\int_0^{\sin^{-1}\frac{a}{c}}\cos\theta\sin\theta\,(a^2 - c^2\sin^2\theta)^{\frac{1}{2}}\,d\theta$$

$$= \frac{4\pi\rho c^6}{(c^2 - a^2)^2}\int_0^{\sin^{-1}\frac{a}{c}}\cos\theta\sin\theta\,(a^2 - c^2\sin^2\theta)^{\frac{1}{2}}\,d\theta$$

$$= \frac{4}{3}\,\frac{\pi\rho c^4}{(c^2 - a^2)^2}\,a^3.$$

Now, if V be the potential at P, we have (see Chap. I. Art. 1)

$$r\frac{d^2(rV)}{dr^2} + \frac{1}{\sin\theta}\frac{d}{d\theta}\left(\sin\theta\frac{dV}{d\theta}\right) = -\frac{4\pi\rho c^5}{r^3}.$$

This is satisfied by $V = -\frac{2}{3}\frac{\pi\rho c^5}{r^3}$.

Assume then, as the complete solution of the equation,

$$V = -\frac{2}{3}\frac{\pi\rho c^5}{r^3} + \left(A_0 + \frac{B_0}{r}\right)P_0 + \left(A_1 r + \frac{B_1}{r^2}\right)P_1 + \ldots$$

$$+ \left(A_i r^i + \frac{B_i}{r^{i+1}}\right)P_i + \ldots.$$

It remains to determine the coefficients A_0, $A_1 \ldots A_i \ldots B_0$, $B_1 \ldots B_i$, so that this expression may not become infinite for any value of r corresponding to a point within the sphere, and that at any point P on the surface of the sphere it may be equal to $\frac{M}{O'P}$, where $O'P : OP :: a : c$, and therefore, at the surface,

$$V = \frac{Mc}{a}\frac{1}{OP} = \frac{4}{3}\frac{\pi\rho c^5 a^2}{(c^2 - a^2)^2 r}.$$

And, at the surface, we have

$$r^2 - 2cr\cos\theta + c^2 - a^2 = 0;$$

$$\therefore \frac{1}{r^3} = \frac{1}{c^2 - a^2}\left(-\frac{1}{r} + \frac{2c\cos\theta}{r^2}\right)$$

$$= \frac{1}{c^2 - a^2}\left(-\frac{P_0}{r} + 2c\frac{P_1}{r^2}\right);$$

$$\therefore \frac{4}{3}\frac{\pi\rho c^5 a^2}{(c^2 - a^2)^2 r} = \frac{2}{3}\frac{\pi\rho c^5}{c^2 - a^2}\left(\frac{P_0}{r} - 2c\frac{P_1}{r^2}\right) + \left(A_0 + \frac{B_0}{r}\right)P_0$$

$$+ \left(A_1 r + \frac{B_1}{r^2}\right)P_1 + \ldots \textit{ identically.}$$

Hence
$$\frac{4}{3}\frac{\pi\rho c^5 a^2}{(c^2 - a^2)^2} = \left(\frac{2}{3}\frac{\pi\rho c^5}{c^2 - a^2} + B_0\right)P_0,$$

$$0 = \left(-\frac{4}{3}\frac{\pi\rho c^6}{c^2 - a^2} + B_1\right)P_1,$$

and B_2, $B_3, \ldots B_i \ldots A_0$, $A_1 \ldots A_i$ all $= 0$.

Hence since $P_0 = 1$,

$$B_0 = \frac{2}{3}\frac{\pi\rho c^5}{c^2 - a^2}\left\{\frac{2a^2}{(c^2 - a^2)} - 1\right\}$$

$$= \frac{2}{3}\frac{\pi\rho c^5}{(c^2 - a^2)^2}(3a^2 - c^2),$$

and

$$B_1 = \frac{4}{3}\frac{\pi\rho c^6}{c^2 - a^2},$$

whence we obtain, as the expression for the potential at any internal point,

$$V = \frac{2}{3}\frac{\pi\rho c^5}{(c^2 - a^2)^2}\frac{3a^2 - c^2}{r} + \frac{4}{3}\frac{\pi\rho c^6}{c^2 - a^2}\frac{\cos\theta}{r^2} - \frac{2}{3}\frac{\pi\rho c^5}{r^3}.$$

6. We shall next proceed to establish the proposition that *if the density of a spherical shell, of indefinitely small thickness, be a zonal surface harmonic, its potential at any internal point will be proportional to the corresponding solid harmonic of positive degree, and its potential at any external point will be proportional to the corresponding solid harmonic of negative degree.*

Take the centre of the sphere as origin, and the axis of the system of zonal harmonics as the axis of z. Let b be the radius of the sphere, δb its thickness, U its volume, so that $U = 4\pi b^2 \delta b$. Let CP_i be the density of the sphere, P_i being the zonal surface harmonic of the degree i, and C any constant.

Draw two planes cutting the sphere perpendicular to the axis of z, at distances from the centre equal to ζ, $\zeta + d\zeta$ respectively. The volume of the strip of the sphere intercepted between these planes will be $\frac{d\zeta}{2b}U$, and its mass will be

$$CP_i \frac{d\zeta}{2b} U.$$

Now $\zeta = b\mu$, hence $d\zeta = b\,d\mu$, and this mass becomes

$$\frac{CU}{2} P_i\, d\mu.$$

Hence the potential of this strip at a point on the axis of z, distant z from the centre, will be

$$\frac{CU}{2}\frac{P_i}{(z^2+b^2-2bz\mu)^{\frac{1}{2}}}d\mu,$$

which may be expanded into

$$\frac{CU}{2b}P_i\left(P_0+P_1\frac{z}{b}+\ldots+P_i\frac{z^i}{b^i}+\ldots\right)d\mu \text{ if } z<b,$$

and $$\frac{CU}{2}\frac{P_i}{z}\left(P_0+P_1\frac{b}{z}+\ldots+P_i\frac{b^i}{z^i}+\ldots\right)d\mu \text{ if } z>b.$$

To obtain the potential of the whole shell, we must integrate these expressions with respect to μ between the limits -1 and $+1$. Hence by the fundamental property of Zonal Harmonics, proved in Chap. II. Art. 10, we get for the potential of the whole shell

$$\frac{CU}{2i+1}\frac{z^i}{b^{i+1}} \text{ at an internal point,}$$

$$\frac{CU}{2i+1}\frac{b^i}{z^{i+1}} \text{ at an external point.}$$

From these expressions for the potential at a point on the axis we deduce, by the method of Art. 1 of the present Chapter, the following expressions for the potential at any point whatever:

$$V_1=\frac{CU}{2i+1}\frac{P_i r^i}{b^{i+1}} \text{ at an internal point,}$$

$$V_2=\frac{CU}{2i+1}\frac{P_i b^i}{r^{i+1}} \text{ at an external point.}$$

From hence we deduce the following expressions for the normal component of the attraction on the point.

Normal component of the attraction on an internal point, measured towards the centre of the sphere,

$$=-\frac{dV_1}{dr}=-\frac{i}{2i+1}CU\frac{P_i r^{i-1}}{b^{i+1}}.$$

Normal component of the attraction on an external point, measured towards the sphere,

$$= -\frac{dV_2}{dr} = \frac{i+1}{2i+1} CU \frac{P_i b^i}{r^{i+2}}.$$

In the immediate neighbourhood of the sphere, where r is indefinitely nearly equal to b, these normal component attractions become respectively

$$-\frac{i}{2i+1} CU \frac{P_i}{b^2}, \quad \frac{i+1}{2i+1} CU \frac{P_i}{b^2},$$

and their difference is therefore

$$CU \frac{P_i}{b^2}.$$

And writing for U its value, $4\pi b^2 \delta b$, this expression becomes

$$4\pi \delta b \,.\, CP_i.$$

Or, the density may be obtained by dividing the algebraic sum of the normal component attractions on two points, one external and the other internal, indefinitely near the sphere, and situated on the same normal, by $4\pi \times$ thickness of the shell.

7. It follows from this that if the density of a spherical shell be expressed by the series

$$C_0P_0 + C_1P_1 + C_2P_2 + \dots + C_iP_i + \dots,$$

$C_0, C_1, C_2 \dots C_i \dots$ being any constants, its potential (V_1) at an internal point will be

$$U\left(\frac{C_0P_0}{b} + \frac{1}{3}\frac{C_1P_1r}{b^2} + \frac{1}{5}\frac{C_2P_2r^2}{b^3} + \dots + \frac{1}{2i+1}\frac{C_iP_ir^i}{b^{i+1}} + \dots\right)$$

and its potential (V_2) at an external point will be

$$U\left(\frac{C_0P_0}{r} + \frac{1}{3}\frac{C_1P_1b}{r^2} + \frac{1}{5}\frac{C_2P_2b^2}{r^3} + \dots + \frac{1}{2i+1}\frac{C_iP_ib^i}{r^{i+1}} + \dots\right).$$

In the last two Articles, by the word "density" is meant "volume density," i.e. the mass of an indefinitely small element of the attracting sphere, divided by the volume of the same element. The product of the volume density of any element of the shell, into the thickness of the shell in the neighbourhood of that element, is called "surface density." We see from the above that, if the surface density be expressed by the series

$$\sigma_0 P_0 + \sigma_1 P_1 + \sigma_2 P_2 + \ldots + \sigma_i P_i + \ldots,$$

the potentials at an internal and an external point will severally be

$$4\pi b^2 \left(\frac{\sigma_0 P_0}{b} + \frac{1}{3}\frac{\sigma_1 P_1 r}{b^2} + \frac{1}{5}\frac{\sigma_2 P_2 r^2}{b^3} + \ldots + \frac{1}{2i+1}\frac{\sigma_i P_i r^i}{b^{i+1}} + \ldots\right),$$

$$4\pi b^2 \left(\frac{\sigma_0 P_0}{r} + \frac{1}{3}\frac{\sigma_1 P_1 b}{r^2} + \frac{1}{5}\frac{\sigma_2 P_2 b^2}{r^3} + \ldots + \frac{1}{2i+1}\frac{\sigma_i P_i b^i}{r^{i+1}} + \ldots\right).$$

This variation in surface density may be obtained either by combining a variable volume density with an uniform thickness, as we have supposed, or by combining a variable thickness with a uniform volume density, or by varying both thickness and density.

8. We have seen, in Chap. II., that any positive integral power of μ, and therefore of course any rational integral function of μ, may be expressed by a finite series of zonal harmonics. It follows, therefore, that we can determine the potential of a spherical shell, whose density is any rational integral function of μ.

Suppose, for instance, we have a shell whose density varies as the square of the distance from a diametral plane. Taking this plane as that of xy, the density may be expressed by $\rho\mu^2$, or $\rho\frac{z^2}{b^2}$. We have seen (Chap. II. Art. 20) that

$$\mu^2 = \frac{1}{3}(1 + 2P_2).$$

Hence, by the result of the last Article, the potential will be

$$\rho\frac{U}{3}\left(\frac{1}{b} + \frac{2}{5}\frac{P_2 r^2}{b^3}\right) \text{ at an internal point,}$$

$$\rho\frac{U}{3}\left(\frac{1}{r}+\frac{2}{5}\frac{P_2 b^2}{r^3}\right) \text{ at an external point;}$$

or, since $P_2 r^2 = \frac{3\mu^2-1}{2} r^2 = \frac{3z^2-r^2}{2}$, we obtain

$$\rho\frac{U}{3}\left(\frac{1}{b}+\frac{1}{5}\frac{3z^2-r^2}{b^3}\right) \text{ for the potential at an internal point,}$$

$$\rho\frac{U}{3}\left\{\frac{1}{r}+\frac{b^2}{5}\left(-\frac{1}{r^3}+\frac{3z^2}{r^5}\right)\right\} \text{ for that at an external point.}$$

9. As an example of the case in which the density is represented by an infinite series of zonal harmonics, suppose we wish to investigate the potential of a spherical shell, whose density varies as the distance from a diameter. Taking this diameter as the axis of z, the density will be represented by $\rho \sin\theta$, or $\rho(1-\mu^2)^{\frac{1}{2}}$. We have investigated in Chap. II. Art. 21, the expansion of $\sin\theta$ in an infinite series of zonal harmonics. Employing this expansion, we shall obtain for the potential

$$\frac{\pi}{2}\frac{\rho U}{b}\left\{\frac{1}{2}P_0 - \frac{1}{16}P_2\frac{r^2}{b^2} - \ldots - \frac{1.3\ldots(i-1)}{2.4\ldots i(i+2)}\cdot\frac{1.3\ldots(i-3)}{2.4\ldots(i-2)i}P_i\frac{r^i}{b^i} - \ldots\right\},$$

or

$$\frac{\pi}{2}\rho U\left\{\frac{1}{2}\frac{P_0}{r} - \frac{1}{16}P_2\frac{b^2}{r^3} - \ldots - \frac{1.3\ldots(i-1)}{2.4\ldots i(i+2)}\cdot\frac{1.3\ldots(i-3)}{2.4\ldots(i-2)i}P_i\frac{b^i}{r^{i+1}} - \ldots\right\},$$

according as the attracted point is internal or external to the spherical shell, i being any even integer. All these expressions may be obtained in terms of surface density, by writing, instead of ρU, $4\pi c^2\sigma$.

10. We may next proceed to shew how the potential of a spherical shell of finite thickness, whose density is any solid zonal harmonic, may be determined. Suppose, for instance, that we have a shell of external radius a, and internal radius a', whose density, at the distance c from the centre, is $\frac{\rho}{h^i}P_i c^i$, h being any line of constant length.

Dividing the sphere into concentric thin spherical shells, of thickness dc, the potential of any one of these shells, of radius c, at an internal point distant r from the centre will be obtained by writing c for b, $\frac{\rho c^i}{h^i}$ for C, $4\pi c^2 dc$ for U, in the first result of Art. 6. This gives

$$\frac{\rho}{h^i}\frac{4\pi c^2 dc}{2i+1}\frac{P_i c^i r^i}{c^{i+1}} \text{ or } \frac{4\pi}{2i+1}\frac{\rho}{h^i}P_i r^i c dc.$$

To obtain the potential of the whole shell, we must integrate this expression, with respect to c, between the limits a' and a. This gives

$$\frac{2\pi}{2i+1}\frac{\rho P_i}{h^i}(a^2-a'^2)\,r^i.$$

Again, the potential of the shell of radius c, at an external point, will be

$$\frac{\rho}{h^i}\frac{4\pi c^2\,dc}{2i+1}\frac{P_i c^{2i}}{r^{i+1}} \text{ or } \frac{4\pi}{2i+1}\frac{\rho}{h^i}P_i\frac{c^{2i+2}}{r^{i+1}}\,dc.$$

Integrating as before, we obtain for the potential of the whole shell,

$$\frac{4\pi}{(2i+1)(2i+3)}\frac{\rho}{h^i}P_i\frac{(a^{2i+3}-a'^{2i+3})}{r^{i+1}}.$$

Suppose now that we wish to obtain the potential of the whole shell at a point forming a part of its mass, distant r from the centre. We shall obtain this by considering separately the two shells into which it may be divided, the external radius of the one, and the internal radius of the other, being each r. Writing r for a', in the first of the foregoing results, we obtain

$$\frac{2\pi}{2i+1}\frac{\rho P_i}{h^i}(a^2-r^2)\,r^i.$$

And writing r for a in the other result, we obtain

$$\frac{4\pi}{(2i+1)(2i+3)}\frac{\rho P_i}{h^i}\frac{r^{2i+3}-a'^{2i+3}}{r^{i+1}}.$$

Adding these, we get for the potential of the whole sphere

$$\frac{4\pi}{2i+1}\frac{\rho P_i}{h^i}\left\{\frac{a^2-r^2}{2}r^i+\frac{r^{2i+3}-a^{2i+3}}{(2i+3)\,r^{i+1}}\right\}.$$

It is hardly necessary to observe that the corresponding results for a solid sphere may be obtained from the foregoing, by putting $a' = 0$.

If the density, instead of being $\frac{\rho}{h^i} P_i c^i$, be $\frac{\rho}{h^m} P_i c^m$, similar reasoning will give us, for the potential of the thin shell of radius c and thickness dc at an internal and external point respectively,

$$\frac{4\pi}{2i+1} \frac{\rho}{h^m} P_i r^i c^{m-i+1} dc, \text{ and } \frac{4\pi}{2i+1} \frac{\rho}{h^m} P_i \frac{c^{i+m+2}}{r^{i+1}} dc.$$

And, integrating as before, we obtain for the potential of the whole shell,

$$\frac{4\pi}{(2i+1)(m-i+2)} \frac{\rho}{h^m} P_i (a^{m-i+2} - a'^{m-i+2}) r^i \text{ at an internal point,}$$

$$\frac{4\pi}{(2i+1)(m+i+3)} \frac{\rho}{h^m} P_i \frac{a^{m+i+3} - a'^{m+i+3}}{r^{i+1}} \text{ at an external point.}$$

And, at a point forming a part of the mass,

$$\frac{4\pi}{2i+1} \frac{\rho P_i}{h^m} \left(\frac{a^{m-i+2} - r^{m-i+2}}{m-i+2} r^i + \frac{r^{m+i+3} - a'^{m+i+3}}{m+i+3} \frac{1}{r^{i+1}} \right).$$

11. Suppose, for example, that we wish to determine, in each of the three cases, the potential of a spherical shell whose external and internal radii are a, a', respectively, and whose density varies as the square of the distance from a diametral plane.

Taking this plane as that of xy, the density may be expressed by $\frac{\rho}{h^2} z^2$, or $\frac{\rho}{h^2} c^2 \mu^2$. Now $\mu^2 = \frac{2P_2 + 1}{3}$. Hence the density of this sphere may be expressed as

$$\frac{2}{3} \frac{\rho}{h^2} P_2 c^2 + \frac{1}{3} \frac{\rho}{h^2} P_0 c^2.$$

The several potentials due to the former term will be, writing 2 for i and multiplying by $\frac{2}{3}$,

$$\frac{4\pi}{15}\frac{\rho P_2}{h^2}(a^2-a'^2)r^2,\ \frac{8\pi}{105}\frac{\rho}{h^2}P_2\frac{a^7-a'^7}{r^3},\ \frac{8\pi}{15}\frac{\rho P_2}{h^2}\left(\frac{a^2-r^2}{2}r^2+\frac{r^7-a'^7}{7r^3}\right).$$

And for the latter term, writing 0 for i, and 2 for m, and multiplying by $\frac{1}{3}$,

$$\frac{4\pi}{12}\frac{\rho}{h^2}(a^4-a'^4),\ \frac{4\pi}{15}\frac{\rho}{h^2}\frac{a^5-a'^5}{r},\ \frac{4\pi}{3}\frac{\rho}{h^2}\left(\frac{a^4-r^4}{4}+\frac{r^5-a'^5}{5r}\right).$$

And, since $P_2r^2=\frac{3z^2-r^2}{2}$, we get for the potential at an internal point

$$\frac{\rho}{h^2}\left\{\frac{2\pi}{15}(a^2-a'^2)(3z^2-r^2)+\frac{\pi}{3}(a^4-a'^4)\right\};$$

at an external point

$$\frac{\rho}{h^2}\left\{\frac{4\pi}{105}\frac{a^7-a'^7}{r^5}(3z^2-r^2)+\frac{4\pi}{15}\frac{a^5-a'^5}{r}\right\};$$

at a point forming a part of the mass

$$\frac{\rho}{h^2}\left\{\frac{4\pi}{15}\left(\frac{a^2-r^2}{2}+\frac{r^7-a'^7}{7r^5}\right)(3z^2-r^2)+\frac{4\pi}{3}\left(\frac{a^4-r^4}{4}+\frac{r^5-a'^5}{5r}\right)\right\}.$$

12. We may now prove that by means of an infinite series of zonal harmonics we may express any function of μ whatever, even a discontinuous function. Suppose, for instance, that we wish to express a function which shall be equal to A from $\mu=1$ to $\mu=\lambda$, and to B from $\mu=\lambda$ to $\mu=-1$. Consider what will be the potential of a spherical shell, radius c, of uniform thickness, whose density is equal to A for the part corresponding to values of μ between 1 and λ, and to B for the part corresponding to values of μ between λ and -1.

Divide the shell, as before, into indefinitely narrow strips by parallel planes, the distance between any two successive planes being $cd\mu$.

We have then, for the potential of such a sphere at any point of the axis, distant z from the centre,

for the first part of the sphere

$$2\pi Ac^2\delta c\int_\lambda^1 \frac{d\mu}{(c^2+z^2-2cz\mu)^{\frac{1}{2}}};$$

and for the latter part

$$2\pi Bc^2\delta c\int_{-1}^\lambda \frac{d\mu}{(c^2+z^2-2cz\mu)^{\frac{1}{2}}}.$$

These are respectively equal to

$$\frac{2\pi Ac^2\delta c}{c}\int_\lambda^1\left(P_0+P_1\frac{z}{c}+P_2\frac{z^2}{c^2}+\ldots+P_i\frac{z^i}{c^i}+\ldots.\right)d\mu,$$

$$\frac{2\pi Bc^2\delta c}{c}\int_{-1}^\lambda\left(P_0+P_1\frac{z}{c}+P_2\frac{z^2}{c^2}+\ldots+P_i\frac{z^i}{c^i}+\ldots.\right)d\mu,$$

at an internal point; and to

$$\frac{2\pi Ac^2\delta c}{z}\int_\lambda^1\left(P_0+P_1\frac{c}{z}+\ldots+P_i\frac{c^i}{z^i}+\ldots.\right)d\mu,$$

$$\frac{2\pi Bc^2\delta c}{z}\int_{-1}^\lambda\left(P_0+P_1\frac{c}{z}+\ldots+P_i\frac{c^i}{z^i}+\ldots.\right)d\mu,$$

at an external point.

Now it follows from Chap. II. (Art. 23) that if i be any positive integer,

$$\int_\lambda^1 P_i d\mu=-\frac{1}{2i+1}\{P_{i+1}(\lambda)-P_{i-1}(\lambda)\},$$

whence, since $\int_{-1}^1 P_i d\mu=0$, it follows that

$$\int_{-1}^\lambda P_i d\mu=\frac{1}{2i+1}\{P_{i+1}(\lambda)-P_{i-1}(\lambda)\}.$$

Also $\int_\lambda^1 P_0 d\mu = 1 - \lambda, \int_{-1}^\lambda P_0 d\mu = 1 + \lambda.$

Hence the above expressions severally become:

For the potential at an internal point on the axis

$$\frac{2\pi c^2 \delta c}{c}\Big[A(1-\lambda) + B(1+\lambda) - \frac{A-B}{3}\{P_2(\lambda) - P_0(\lambda)\}\frac{z}{c}$$
$$-\frac{A-B}{5}\{P_3(\lambda) - P_1(\lambda)\}\frac{z^2}{c^2} - \dots$$
$$-\frac{A-B}{2i+1}\{P_{i+1}(\lambda) - P_{i-1}(\lambda)\}\frac{z^i}{c^i} - \dots\Big];$$

and for the potential at an external point on the axis

$$2\pi c^2 \delta c\Big[\frac{A(1-\lambda) + B(1+\lambda)}{z} - \frac{A-B}{3}\{P_2(\lambda) - P_0(\lambda)\}\frac{c}{z^2}$$
$$-\frac{A-B}{5}\{P_3(\lambda) - P_1(\lambda)\}\frac{c^2}{z^3} - \dots$$
$$-\frac{A-B}{2i+1}\{P_{i+1}(\lambda) - P_{i-1}(\lambda)\}\frac{c^i}{z^{i+1}} - \dots\Big].$$

Hence the potentials at a point situated anywhere are respectively

$$\frac{2\pi c^2 \delta c}{c}\Big[\{A(1-\lambda) + B(1+\lambda)\}P_0(\mu)$$
$$-\frac{A-B}{3}\{P_2(\lambda) - P_0(\lambda)\}\frac{P_1(\mu) \cdot r}{c}$$
$$-\frac{A-B}{5}\{P_3(\lambda) - P_1(\lambda)\}\frac{P_2(\mu) \cdot r^2}{c^2} - \dots$$
$$-\frac{A-B}{2i+1}\{P_{i+1}(\lambda) - P_{i-1}(\lambda)\}\frac{P_i(\mu) r^i}{c^i} - \dots\Big],$$

at an internal point;

and

$$2\pi c^2\delta c\left[\{A(1-\lambda)+B(1+\lambda)\}\frac{P_0(\mu)}{r}\right.$$

$$-\frac{A-B}{3}\{P_2(\lambda)-P_0(\lambda)\}\frac{P_1(\mu)c}{r^2}$$

$$-\frac{A-B}{5}\{P_3(\lambda)-P_1(\lambda)\}\frac{P_2(\mu)c^2}{r^3}-\ldots\ldots$$

$$\left.-\frac{A-B}{2i+1}\{P_{i+1}(\lambda)-P_{i-1}(\lambda)\}\frac{P_i(\mu)c^i}{r^{i+1}}-\ldots\ldots\right]$$

at an external point.

Now, if we inquire what will be the potential for the following distribution of density,

$$\tfrac{1}{2}[A(1-\lambda)+B(1+\lambda)-(A-B)\{P_2(\lambda)-P_0(\lambda)\}P_1(\mu)$$
$$-(A-B)\{P_3(\lambda)-P_1(\lambda)\}P_2(\mu)-\ldots$$
$$-(A-B)\{P_{i+1}(\lambda)-P_{i-1}(\lambda)\}P_i(\mu)-\ldots],$$

we see by Art. 6 that it will be exactly the same, both at an internal and for an external point, as that above investigated for the shell made up of two parts, whose densities are A and B respectively.

But it is known that there is one, and only one, distribution of attracting matter over a given surface, which will produce a specified potential at every point, both external and internal. Hence the above expression must represent exactly the same distribution of density. That is, writing the above series in a slightly different form, the expression

$$\frac{A+B}{2}-\frac{A-B}{2}\left[\lambda+\{P_2(\lambda)-P_0(\lambda)\}P_1(\mu)\right.$$
$$+\{P_3(\lambda)-P_1(\lambda)\}P_2(\mu)+\ldots$$
$$\left.\ldots+\{P_{i+1}(\lambda)-P_{i-1}(\lambda)\}P_i(\mu)+\ldots\right]$$

is equal to A, for all values of μ from 1 to λ, and to B for all values of μ from λ to -1.

13. By a similar process, any other discontinuous function, whose values are given for all values of μ from 1 to -1, may be expressed. Suppose, for instance, we wish to express a function which is equal to A from $\mu = 1$ to $\mu = \lambda_1$, to B from $\mu = \lambda_1$ to $\mu = \lambda_2$, and to C from $\mu = \lambda_2$ to $\mu = -1$. This will be obtained by adding the two series

$$\frac{A-B}{2} - \frac{A-B}{2}[\lambda_1 + \{P_2(\lambda_1) - P_0(\lambda_1)\}P_1(\mu) + \dots$$
$$+ \{P_{i+1}(\lambda_1) - P_{i-1}(\lambda_1)\}P_i(\mu) + \dots],$$

$$\frac{B+C}{2} - \frac{B-C}{2}[\lambda_2 + \{P_2(\lambda_2) - P_0(\lambda_2)\}P_1(\mu) + \dots$$
$$+ \{P_{i+1}(\lambda_2) - P_{i-1}(\lambda_2)\}P_i(\mu) + \dots].$$

For the former is equal to $A - B$ from $\mu = 1$ to $\mu = \lambda_1$, and to 0 from $\mu = \lambda_1$ to $\mu = -1$; and the latter is equal to B from $\mu = 1$ to $\mu = \lambda_2$, and to C from $\mu = \lambda_2$ to $\mu = -1$.

By supposing A and C each $= 0$, and $B = 1$, we deduce a series which is equal to 1 for all values of μ from $\mu = \lambda_1$ to $\mu = \lambda_2$, and zero for all other values. This will be

$$\frac{1}{2}[\lambda_1 - \lambda_2 + \{\overline{P_2(\lambda_1) - P_2(\lambda_2)} - \overline{P_0(\lambda_1) - P_0(\lambda_2)}\}P_1(\mu) + \dots$$
$$+ \{\overline{P_{i+1}(\lambda_1) - P_{i+1}(\lambda_2)} - \overline{P_{i-1}(\lambda_1) - P_{i-1}(\lambda_2)}\}P_i(\mu) + \dots].$$

This may be verified by direct investigation of the potential of the portion of a homogeneous spherical shell, of density unity, comprised between two parallel planes, distant respectively $c\lambda_1$ and $c\lambda_2$ from the centre of the spherical shell.

14. In the case in which λ_1 and λ_2 are indefinitely nearly equal to each other, let $\lambda_2 = \lambda$, and $\lambda_1 = \lambda + d\lambda$. We then have, ultimately,

$$P_i(\lambda_1) - P_i(\lambda_2) = \frac{dP_i(\lambda)}{d\lambda}\, d\lambda.$$

Hence $\overline{P_{i+1}(\lambda_1) - P_{i+1}(\lambda_2)} - \overline{P_{i-1}(\lambda_1) - P_{i-1}(\lambda_2)}$

$$= \left\{\frac{dP_{i+1}(\lambda)}{d\lambda} - \frac{dP_{i-1}(\lambda)}{d\lambda}\right\} d\lambda$$

$$= (2i+1)P_i(\lambda)d\lambda.$$

Hence the series

$$\frac{d\lambda}{2}\{1 + 3P_1(\lambda)P_1(\mu) + 5P_2(\lambda)P_2(\mu) + \ldots$$
$$+ (2i+1)\,P_i(\lambda)P_i(\mu) + \ldots\}$$

is equal to 1 when $\mu = \lambda$ (or, more strictly, when μ has any value from λ to $\lambda + d\lambda$) and is equal to 0 for all other values of μ.

We hence infer that

$$1 + 3P_1(\lambda)P_1(\mu) + \ldots + (2i+1)P_i(\lambda)P_i(\mu) + \ldots$$

is infinite when $\mu = \lambda$, and zero for all other values of μ.

15. Representing the series

$$\tfrac{1}{2}\{1 + 3P_1(\lambda)P_1(\mu) + \ldots + (2i+1)P_i(\lambda)P_i(\mu) + \ldots\}$$

by $\phi(\lambda)$ for the moment, we see that $\rho\phi(\lambda)d\lambda$ is equal to ρ when $\mu = \lambda$, and to zero for all other values. Hence the expression

$$\{\rho_1\phi(\lambda_1) + \rho_2\phi(\lambda_2) + \ldots\}d\lambda$$

is equal to ρ_1 when $\mu = \lambda_1$, to ρ_2 when $\mu = \lambda_2 \ldots$ Supposing now that $\lambda_1, \lambda_2 \ldots$ are a series of values varying continuously from 1 to -1, we see that this expression becomes

$$\int_{-1}^{1} \rho\phi(\lambda)d\lambda,$$

ρ being any function of λ, continuous or discontinuous. Hence, writing $\phi(\lambda)$ at length, we see that

$$\frac{1}{2}\left\{\int_{-1}^{1} \rho d\lambda + 3P_1(\mu)\int_{-1}^{1} \rho P_1(\lambda)d\lambda + \ldots\right.$$
$$\left. + (2i+1)P_i(\mu)\int_{-1}^{1} \rho P_i(\lambda)d\lambda + \ldots\right\}$$

is equal, for all values of μ from -1 to $+1$, to the same function of μ that ρ is of λ.

16. The same conclusion may be arrived at as follows:

The potential of a spherical shell, whose density is ρ, and volume U, at any point on the axis of z, is

$$\frac{U}{2}\int_{-1}^{1}\frac{\rho d\lambda}{(c^2+z^2-2\lambda cz)^{\frac{1}{2}}},$$

which is equal to $\frac{U}{2c}\Big\{\int_{-1}^{1}\rho d\lambda+\frac{z}{c}\int_{-1}^{1}\rho P_1(\lambda)d\lambda+\dots$

$$+\frac{z^i}{c^i}\int_{-1}^{1}\rho P_i(\lambda)d\lambda+\dots\Big\},$$

for an internal point,

and to $\frac{U}{2}\Big\{\frac{1}{z}\int_{-1}^{1}\rho d\lambda+\frac{c}{z^2}\int_{-1}^{1}\rho P_1(\lambda)d\lambda+\dots$

$$+\frac{c^i}{z^{i+1}}\int_{-1}^{1}\rho P_i(\lambda)d\lambda+\dots\Big\},$$

for an external point.

It hence follows that the potential, at a point situated anywhere, is

$$\frac{U}{2c}\Big\{\int_{-1}^{1}\rho d\lambda+\frac{P_1(\mu)\,r}{c}\int_{-1}^{1}\rho P_1(\lambda)d\lambda+\dots$$

$$+\frac{P_i(\mu)r^i}{c^i}\int_{-1}^{1}\rho P_i(\lambda)d\lambda+\dots\Big\},$$

for an internal point,

and to $\frac{U}{2}\Big\{\frac{1}{r}\int_{-1}^{1}\rho d\lambda+\frac{P_1(\mu)c}{r^2}\int_{-1}^{1}\rho P_1(\lambda)d\lambda+\dots$

$$+\frac{P_i(\mu)c^i}{r^{i+1}}\int_{-1}^{1}\rho P_i(\lambda)d\lambda+\dots\Big\},$$

for an external point.

And these expressions are respectively equal to those for the potentials, at an internal and external point respectively, for matter distributed according to the following law of density:

$$\frac{1}{2}\left\{\int_{-1}^{1}\rho d\lambda + 3P_1(\mu)\int_{-1}^{1}\rho P_1(\lambda)d\lambda + \dots \right.$$
$$\left. + (2i+1)P_i(\mu)\int_{-1}^{1}\rho P_i(\lambda)d\lambda + \dots\right\}.$$

It will be observed, in applying this formula, that if ρ be a discontinuous function of λ, each of the expressions of the form $\int_{-1}^{1}\rho P_i(\lambda)d\lambda$ will be the sum of the results of a series of integrations, each integration being taken through a series of values of λ, for which ρ varies continuously.

CHAPTER IV.

SPHERICAL HARMONICS IN GENERAL. TESSERAL AND SECTORIAL HARMONICS. ZONAL HARMONICS WITH THEIR AXIS IN ANY POSITION. POTENTIAL OF A SOLID NEARLY SPHERICAL IN FORM.

1. WE have hitherto discussed those solutions of the equation $\nabla^2 V=0$ which are symmetrical about the axis of z, or in other words, those solutions of the equivalent equation in polar co-ordinates which are independent of ϕ. We propose, in the present Chapter, to consider the forms of spherical harmonics in general, understanding by a Solid Spherical Harmonic of the ith degree a rational integral homogeneous function of x, y, z, of the ith degree which satisfies the equation $\nabla^2 V=0$, and by a Surface Spherical Harmonic of the ith degree the quotient obtained by dividing a Solid Spherical Harmonic by $(x^2+y^2+z^2)^{\frac{i}{2}}$. Such an expression, as we see by writing $x=r\sin\theta\cos\phi$, $y=r\sin\theta\sin\phi$, $z=r\cos\theta$, will be of the ith degree in $\sin\theta\cos\phi$, $\sin\theta\sin\phi$, $\cos\theta$; and will satisfy the differential equation in Y_i,

$$\frac{1}{\sin\theta}\frac{d}{d\theta}\left(\sin\theta\frac{dY_i}{d\theta}\right)+\frac{1}{\sin^2\theta}\frac{d^2Y_i}{d\phi^2}+i(i+1)Y_i=0,$$

or, writing μ for $\cos\theta$,

$$\frac{d}{d\mu}\left\{(1-\mu^2)\frac{dY_i}{d\mu}\right\}+\frac{1}{1-\mu^2}\frac{d^2Y_i}{d\phi^2}+i(i+1)Y_i=0.$$

It will be convenient, before proceeding to investigate the algebraical forms of these expressions, to discuss some of their simpler physical properties.

2. We will then proceed to shew how spherical harmonics may be employed to determine the potential, and

consequently the attraction, of a spherical shell of indefinitely small thickness.

We will first establish an important theorem, connecting the potential of such a shell on an external point with that on a corresponding internal point. The theorem is as follows:

If O *be the centre of such a shell,* c *its radius,* P *any internal point,* P′ *an external point, so situated that* P′ *lies on* OP *produced, and that* OP . OP′ = c², *and if* OP = r, OP′ = r′, *then the potential of the shell at* P *is to its potential at* P′ *as* c *to* r, *or (which is the same thing) as* r′ *to* c.

For, let A be the point where OP' meets the surface of the sphere, Q any other point of its surface. Then, by a known geometrical theorem,

$$QP : QP' :: AP : AP' :: c - r : r' - c.$$

And $$\frac{c-r}{r'-c} = \frac{cr - r^2}{rr' - cr} = \frac{cr - r^2}{c^2 - cr} = \frac{r}{c} = \frac{c}{r'}.$$

Again, considering the element of the shell in the immediate neighbourhood of Q, its potential at P is to its potential at P' as QP' is to QP, that is, as c to r, or (which is the same thing) as r' to c, which ratio, being independent of the position of Q, must be true for every element of the spherical shell, and therefore for the whole shell. Hence the proposition is proved.

3. Now, suppose the law of density of the shell to be such that its potential at any internal point is $F(\mu, \phi)\,\frac{r^i}{c^i}$.

Then $F(\mu, \phi)\,\frac{r^i}{c^i}$ must be a solid harmonic of the degree i.

Hence $F(\mu, \phi)$ must be a surface harmonic of the degree i. Let us represent it by Y_i.

By the proposition just proved, the potential at any external point, distant r' from the centre, must be

$$Y_i\,\frac{c^{i+1}}{r'^{i+1}}.$$

Hence, the component of the attraction of the sphere on the internal point measured in the direction from the point inwards, i.e. towards the centre of the sphere, is

$$-i\, Y_i \frac{r^{i-1}}{c^i}.$$

And the component in the same direction of the attraction on the external point, measured inwards, is

$$(i+1)\, Y_i \frac{c^i}{r'^{i+1}}.$$

Now suppose the two points to lie on the same line passing through the centre of the sphere, and to be both indefinitely close to the surface of the sphere, so that r and r' are each indefinitely nearly equal to c.

And the attraction on the external point exceeds the attraction on the internal point by

$$(2i+1)\frac{Y_i}{c}.$$

Now, supposing the shell to be divided into two parts, by a plane passing through the internal point perpendicular to the line joining it with the centre, we see that the attraction of the larger part of the shell on the two points will be ultimately the same, while the component attractions of the smaller portions, in the direction above considered, will be equal in magnitude and opposite in direction. Hence the difference between these components, viz. $(2i+1)\frac{Y_i}{c}$, will be equal to twice the component attraction of the smaller portion in the direction of the line joining the two points. But if ρ_i be the density of the shell, δc its thickness, this component attraction is $2\pi\rho_i\delta c$.

Hence $$(2i+1)\frac{Y_i}{c} = 4\pi\rho_i\delta c,$$

or $$\rho_i = \frac{2i+1}{4\pi c\delta c} Y_i.$$

And, if σ_i be the corresponding surface density,

$$\sigma_i = \frac{2i+1}{4c} Y_i.$$

It hence follows that *if the potential of a spherical shell, of indefinitely small thickness, be a surface harmonic, its potential at any internal point will be proportional to the corresponding solid harmonic of positive degree, and its potential at any external point will be proportional to the corresponding solid harmonic of negative degree.*

That is, the proposition proved for zonal harmonics in Chap. III. Art. 6, is now extended to spherical harmonics in general.

4. *The spherical harmonic of the degree* i *will involve* 2i + 1 *arbitrary constants.*

For the solid spherical harmonic, $r^i Y_i$, being a rational integral function of x, y, z of the ith degree, will consist of $\frac{(i+1)(i+2)}{2}$ terms. Now the expression $\nabla^2 V$, being a rational integral function of x, y, z of the degree $i-2$, will consist of $\frac{(i-1)i}{2}$ terms; and the condition that it must be $=0$ for all values of x, y, z, will give rise to $\frac{(i-1)i}{2}$ relations among the $\frac{(i+1)(i+2)}{2}$ coefficients of these terms, leaving $\frac{(i+1)(i+2)}{2} - \frac{(i-1)i}{2}$, or $2i+1$, independent coefficients.

5. We proceed to shew how the spherical harmonic of the degree i may be arranged in a series of terms, each of which may be deduced by differentiation from the Zonal Harmonic symmetrical about the axis of z. The solid zonal harmonic, which, in accordance with the notation already employed, is represented by $r^i P_i(\mu)$, is a function of z and r of the degree i, satisfying the equation $\nabla^2 V = 0$, or $\frac{d^2V}{dx^2} + \frac{d^2V}{dy^2} + \frac{d^2V}{dz^2} = 0$.

Now, if we denote this expression by $P_i(z)$, we see that

since it is a function of z and r, it is a function of the distance (z) from a certain plane passing through the origin, and of the distance (r) from the origin. Further, if we write for z the distance from any other plane passing through the origin, leaving r unaltered, the equation $\frac{d^2V}{dx^2}+\frac{d^2V}{dy^2}+\frac{d^2V}{dz^2}=0$ will continue to be satisfied.

Now $z+\alpha(x+\sqrt{-1}\,y)$, α being any quantity whatever, represents the distance from a certain plane passing through the origin, since in this expression, the sum of the squares of the coefficients of z, x, y is equal to unity. Hence $P_i\{z+\alpha(x+\sqrt{-1}\,y)\}$ is a solid zonal harmonic of the degree i, its axis being the imaginary line $\frac{x}{\alpha}=\frac{y}{\alpha\sqrt{-1}}=z$. Therefore the equation

$$\frac{d^2V}{dx^2}+\frac{d^2V}{dy^2}+\frac{d^2V}{dz^2}=0,$$

is satisfied by $V=P_i\{z+\alpha(x+\sqrt{-1}\,y)\}$, that is, expanding by Taylor's Theorem, it is satisfied by

$$P_i(z)+\alpha(x+\sqrt{-1}\,y)\frac{dP_i(z)}{dz}+\frac{\alpha^2}{1.2}(x+\sqrt{-1}\,y)^2\frac{d^2P_i(z)}{dz^2}+\dots$$
$$+\frac{\alpha^i(x+\sqrt{-1}\,y)^i}{1.2\dots i}\frac{d^iP_i(z)}{dz^i},$$

for all values of α.

Hence, since the equation in V is linear, it follows that it is satisfied by each term separately, or that, besides $P_i(z)$ itself, each of the i expressions,

$$(x+\sqrt{-1}\,y)\frac{dP_i(z)}{dz},\ (x+\sqrt{-1}\,y)^2\frac{d^2P_i(z)}{dz^2},\ \dots(x+\sqrt{-1}\,y)^i\frac{d^iP_i(z)}{dz^i},$$

satisfies the equation $V=0$.

By similar reasoning we may shew that each of the i expressions,

$$(x-\sqrt{-1}\,y)\frac{dP_i(z)}{dz},\ (x-\sqrt{-1}\,y)^2\frac{d^2P_i(z)}{dz^2},\ \dots(x-\sqrt{-1}\,y)^i\frac{d^iP_i(z)}{dz^i},$$

satisfies the same equation.

Now each of the $2i$ solutions, thus obtained, is imaginary. But the sum of any two or more of them, or the result obtained by multiplying any two or more by any arbitrary quantities, and adding the results together, will also be a solution of the equation. Hence, adding each term of the first series to the corresponding term of the second, we obtain a series of i real solutions of the equation. Another such series may be obtained by subtracting each term of the second series from the corresponding term of the first, and dividing by $\sqrt{-1}$. We have thus obtained (including the original term $P_i(z)$) a series of $2i+1$ independent solutions of the given equation, which will be the $2i+1$ independent solid harmonics of the degree i.

6. We may deduce the surface harmonics from these by writing $r\sin\theta\cos\phi$ for x, $r\sin\theta\sin\phi$ for y, $r\cos\theta$ for z, and dividing by r^i. Then, putting $\cos\theta = \mu$, and observing that $P_i(z) = r^i P_i(\mu)$, $\frac{dP_i(z)}{dz} = r^i \frac{dP_i(\mu)}{d\mu} \ldots$ we obtain the following series of $2i+1$ solutions:

$$P_i(\mu),$$

$$\cos\phi\sin\theta\frac{dP_i(\mu)}{d\mu},\ \cos 2\phi\sin^2\theta\frac{d^2P_i(\mu)}{d\mu^2},\ \ldots\cos i\phi\sin^i\theta\frac{d^iP_i(\mu)}{d\mu^i},$$

$$\sin\phi\sin\theta\frac{dP_i(\mu)}{d\mu},\ \sin 2\phi\sin^2\theta\frac{d^2P_i(\mu)}{d\mu^2},\ \ldots\sin i\phi\sin^i\theta\frac{d^iP_i(\mu)}{d\mu^i}.$$

Expressions of the form

$$C\cos\sigma\phi\sin^\sigma\theta\frac{d^\sigma P_i(\mu)}{d\mu^\sigma},$$

or

$$S\sin\sigma\phi\sin^\sigma\theta\frac{d^\sigma P_i(\mu)}{d\mu^\sigma},$$

or their equivalents,

$$C\cos\sigma\phi\,(1-\mu^2)^{\frac{\sigma}{2}}\frac{d^\sigma P_i(\mu)}{d\mu^\sigma},$$

$$S\sin\sigma\phi\,(1-\mu^2)^{\frac{\sigma}{2}}\frac{d^\sigma P_i(\mu)}{d\mu^\sigma},$$

(C and S denoting any quantities independent of θ and ϕ) are called Tesseral Surface Harmonics of the degree i and order σ. The particular forms assumed by them when $\sigma = i$ are called Sectorial Surface Harmonics of the degree i. It will be observed that, since $\frac{d^i P_i(\mu)}{d\mu^i}$ is a numerical constant, Sectorial Harmonics only involve θ in the form

$$\sin^i\theta, \text{ or } (1-\mu^2)^{\frac{i}{2}}.$$

The product obtained by multiplying a Tesseral or Sectorial Surface Harmonic of the degree i by r^i (that is, the expression directly obtained in Art. 5) is called a Tesseral or Sectorial Solid Harmonic of the degree i.

7. We shall denote the factor of a Tesseral or Sectorial Harmonic which is a function of θ, that is $\sin^\sigma\theta \frac{d^\sigma P_i(\mu)}{d\mu^\sigma}$, or $(1-\mu^2)^{\frac{\sigma}{2}} \frac{d^\sigma P_i(\mu)}{d\mu^\sigma}$, by the symbol $T_i^{(\sigma)}$, or, when it is necessary to particularize the quantity of which it is a function, by $T_i^{(\sigma)}(\mu)$ or $T_i^{(\sigma)}(\cos\theta)$.

It will be convenient, for the purpose of comparison with the forms of Tesseral Harmonics given in the *Mécanique Céleste*, and elsewhere, to obtain $T_i^{(\sigma)}$ in a completely developed form.

Now, since $P_i(\mu) = \frac{1}{2^i.1.2.3\ldots i} \cdot \frac{d^i(\mu^2-1)^i}{d\mu^i}$, we see that

$$\frac{d^\sigma P_i(\mu)}{d\mu^\sigma} = \frac{1}{2^i.1.2.3\ldots i} \cdot \frac{d^{i+\sigma}(\mu^2-1)^i}{d\mu^{i+\sigma}}$$

$$= \frac{1}{2^i.1.2.3\ldots i} \frac{d^{i+\sigma}}{d\mu^{i+\sigma}} \left\{\mu^{2i} - \frac{i}{1}\mu^{2i-2} + \frac{i(i-1)}{1.2}\mu^{2i-4} - \ldots\right\}.$$

Now $\frac{d^{i+\sigma}}{d\mu^{i+\sigma}} \left\{\mu^{2i} - \frac{i}{1}\mu^{2i-2} + \frac{i(i-1)}{1.2}\mu^{2i-4} - \ldots\right\}$

$= 2i(2i-1)\ldots(i-\sigma+1)\,\mu^{i-\sigma}$

$$-\frac{i}{1}(2i-2)(2i-3)\dots(i-\sigma-1)\,\mu^{i-\sigma-2}$$

$$+\frac{i(i-1)}{1\,.\,2}(2i-4)(2i-5)\dots(i-\sigma-3)\mu^{i-\sigma-4}$$

$$-\dots\dots$$

$$=2i(2i-1)\dots(i-\sigma+1)\left\{\mu^{i-\sigma}-\frac{(i-\sigma)(i-\sigma-1)}{2\,.\,2i-1}\mu^{i-\sigma-2}\right.$$

$$\left.+\frac{(i-\sigma)(i-\sigma-1)(i-\sigma-2)(i-\sigma-3)}{2\,.\,4\,.(2i-1)(2i-3)}\mu^{i-\sigma-4}-\dots\right\}.$$

And therefore

$$T_i^{(\sigma)}=\frac{2i(2i-1)\dots(i-\sigma+1)}{2^i\,.\,1\,.\,2\,.\,3\dots i}(1-\mu^2)^{\frac{\sigma}{2}}\left\{\mu^{i-\sigma}-\frac{(i-\sigma)(i-\sigma-1)}{2(2i-1)}\mu^{i-\sigma-2}\right.$$

$$\left.+\frac{(i-\sigma)(i-\sigma-1)(i-\sigma-2)(i-\sigma-3)}{2.4\,(2i-1)(2i-3)}\mu^{i-\sigma-4}-\dots\right\}.$$

The form given by Laplace for a Tesseral Surface Harmonic of the degree i and order σ is (see *Mécanique Céleste,* Liv. 3, Chap. 2, pp. 40—47)

$$A(1-\mu^2)^{\frac{\sigma}{2}}\left\{\mu^{i-\sigma}-\frac{(i-\sigma)(i-\sigma-1)}{2(2i-1)}\mu^{i-\sigma-2}+\dots\right\}\cos\sigma\phi,$$

A being a quantity independent of θ and ϕ. The factor of this, involving μ, is denoted by Thomson and Tait (*Natural Philosophy*, Vol. 1, p. 149) by the symbol $\Theta_i^{(\sigma)}$. Thomson and Tait also employ a symbol $\vartheta_i^{(\sigma)}$, adopted by Maxwell in his Treatise on Electricity and Magnetism, Vol. 1, p. 164, which is equal to

$$2^\sigma\frac{1\,.\,2\dots\sigma}{(i+\sigma)(i+\sigma-1)\dots(i-\sigma+1)}(1-\mu^2)^{\frac{\sigma}{2}}\frac{d^\sigma P_i(\mu)}{d\mu^\sigma},$$

or $$2^\sigma\frac{1\,.\,2\dots\sigma}{(i+\sigma)(i+\sigma-1)\dots(i-\sigma+1)}T_i^{(\sigma)}.$$

Heine represents the expression

$$(\mu^2-1)^{\frac{\sigma}{2}}\left\{\mu^{i-\sigma}-\frac{(i-\sigma)(i-\sigma-1)}{2(2i-1)}\mu^{i-\sigma-2}\right.$$
$$\left.+\frac{(i-\sigma)(i-\sigma-1)(i-\sigma-2)(i-\sigma-3)}{2.4.(2i-1)(2i-3)}\mu^{i-\sigma-4}-\ldots\right\},$$

or $(-1)^{\frac{\sigma}{2}}\Theta_i^{(\sigma)}$, by the symbol $P_\sigma^i(\mu)$, and calls these expressions by the name *Zugeordnete Functionen Erster Art* (*Handbuch der Kugelfunctionen*, pp. 117, 118) which Todhunter translates by the term "Associated Functions of the First Kind," which we shall adopt.

Heine also represents the series

$$\mu^{i-\sigma}-\frac{(i-\sigma)(i-\sigma-1)}{2(2i-1)}\mu^{i-\sigma-2}$$
$$+\frac{(i-\sigma)(i-\sigma-1)(i-\sigma-2)(i-\sigma-3)}{2.4(2i-1)(2i-3)}\mu^{i-\sigma-4},$$

by the symbol $\mathfrak{P}_\sigma^i(\mu)$, (p. 117).

The several expressions, $T_i^{(\sigma)}$, $\Theta_i^{(\sigma)}$, $\vartheta_i^{(\sigma)}$, P_σ^i, $\mathfrak{P}_\sigma^i$, are connected together as follows:

$$\frac{2^i.1.2.3\ldots i}{2i(2i-1)\ldots(i-\sigma+1)}T_i^{(\sigma)}=\Theta_i^{(\sigma)}$$
$$=\frac{2^{i-\sigma}i(i-1)\ldots(\sigma+1)}{(i+\sigma+1)(i+\sigma+2)\ldots 2i}\vartheta_i^{(\sigma)}=(-1)^{\frac{\sigma}{2}}P_\sigma^i=(1-\mu^2)^{\frac{\sigma}{2}}\mathfrak{P}_\sigma^i.$$

8. It has been already remarked that the roots of the equation $P_i=0$ are all real. It follows also that those of the equations $\frac{dP_i}{d\mu}=0$, $\frac{d^2P}{d\mu^2}=0\ldots$ are real also. Hence we may arrive at the following conclusions, concerning the curves, traced on a sphere, which result from our putting any one of these series of spherical harmonics $=0$.

By putting a zonal harmonic$=0$, we obtain i small circles, whose planes are parallel to one another, perpendicular to the axis of the zonal harmonic, and symmetrically situated with respect to the diametral plane, perpendicular to this axis. If i be an odd number this diametral plane itself becomes one of the series.

By putting the tesseral harmonic of the order $\sigma=0$, we obtain $i-\sigma$ small circles, situated as before, and σ great circles, determined by the equation $\cos\sigma\phi=0$, or $\sin\sigma\phi=0$, as the case may be, their planes all intersecting in the axis of the system of harmonics, the angle between the planes of any two consecutive great circles being $\frac{\pi}{\sigma}$.

By putting the sectorial harmonic $=0$, we obtain i great circles, whose planes all intersect in the axis of the system, the angle between any two consecutive planes being $\frac{\pi}{i}$.

9. The tesseral harmonic may be regarded from another point of view. Suppose it is required to determine a solid harmonic of the degree i, and of the form $Y_i r^i$, such that Y_i shall be the product of a function of μ, and of a function of ϕ, which functions we will denote by the symbols M_i, Φ_i, respectively. The differential equation, to which this will lead, is

$$i(i+1)\,M_i\Phi_i+\frac{d}{d\mu}\left\{(1-\mu^2)\frac{dM_i}{d\mu}\right\}\Phi_i+\frac{M_i}{1-\mu^2}\frac{d^2\Phi_i}{d\phi^2}=0.$$

Now this will be satisfied, if we make M_i and Φ_i satisfy the following two equations:

$$i(i+1)\,M_i+\frac{d}{d\mu}\left\{(1-\mu^2)\frac{dM_i}{d\mu}\right\}=\frac{\sigma^2}{1-\mu^2}M_i,$$

$$\frac{d^2\Phi_i}{d\phi^2}=-\sigma^2\Phi_i.$$

The latter equation gives

$$\Phi_i=C\cos\sigma\phi+C'\sin\sigma\phi.$$

And, taking σ as an integer, positive or negative, the former is satisfied by $M_i = T_i^{(\sigma)}$, i.e. $(1-\mu^2)^{\frac{\sigma}{2}}\left(\frac{d}{d\mu}\right)^{i+\sigma}(1-\mu^2)^i$, as we proceed to prove.

We know that

$$\frac{d}{d\mu}\left\{(1-\mu^2)\frac{dP_i}{d\mu}\right\}+i(i+1)P_i=0.$$

Differentiate σ times, and we get

$$\frac{d^{\sigma+1}}{d\mu^{\sigma+1}}\left\{(1-\mu^2)\frac{dP_i}{d\mu}\right\}+i(i+1)\frac{d^\sigma P_i}{d\mu^\sigma}=0;$$

whence, by Leibnitz's Theorem,

$$(1-\mu^2)\frac{d^{\sigma+2}P_i}{d\mu^{\sigma+2}}-2(\sigma+1)\mu\frac{d^{\sigma+1}P_i}{d\mu^{\sigma+1}}-(\sigma+1)\sigma\frac{d^\sigma P_i}{d\mu^\sigma}$$
$$+i(i+1)\frac{d^\sigma P_i}{d\mu^\sigma}=0,$$

or

$$(1-\mu^2)\frac{d^{\sigma+2}P_i}{d\mu^{\sigma+2}}-2(\sigma+1)\mu\frac{d^{\sigma+1}P_i}{d\mu^{\sigma+1}}+(i-\sigma)(i+\sigma+1)\frac{d^\sigma P_i}{d\mu^\sigma}=0,$$

and, multiplying by $(1-\mu^2)^{\frac{\sigma}{2}}$,

$$(1-\mu^2)^{\frac{\sigma}{2}+1}\frac{d^{\sigma+2}P_i}{d\mu^{\sigma+2}}-2(\sigma+1)\mu(1-\mu^2)^{\frac{\sigma}{2}}\frac{d^{\sigma+1}P_i}{d\mu^{\sigma+1}}$$
$$+(i-\sigma)(i+\sigma+1)(1-\mu^2)^{\frac{\sigma}{2}}\frac{d^\sigma P_i}{d\mu^\sigma}=0 \ldots (1).$$

Now, putting $\quad (1-\mu^2)^{\frac{\sigma}{2}}\frac{d^\sigma P_i}{d\mu^\sigma}=T_i^{(\sigma)}$,

we get

$$\frac{dT_i^{(\sigma)}}{d\mu}=(1-\mu^2)^{\frac{\sigma}{2}}\frac{d^{\sigma+1}P_i}{d\mu^{\sigma+1}}-\sigma\mu(1-\mu^2)^{\frac{\sigma}{2}-1}\frac{d^\sigma P_i}{d\mu^\sigma};$$

$$\therefore\ (1-\mu^2)\frac{dT_i^{(\sigma)}}{d\mu}=(1-\mu^2)^{\frac{\sigma}{2}+1}\frac{d^{\sigma+1}P_i}{d\mu^{\sigma+1}}-\sigma\mu(1-\mu^2)^{\frac{\sigma}{2}}\frac{d^\sigma P_i}{d\mu^\sigma},$$

$$\frac{d}{d\mu}\left\{(1-\mu^2)\frac{dT_i^{(\sigma)}}{d\mu}\right\} = (1-\mu^2)^{\frac{\sigma}{2}+1}\frac{d^{\sigma+2}P_i}{d\mu^{\sigma+2}}$$

$$-2(\sigma+1)\mu(1-\mu^2)^{\frac{\sigma}{2}}\frac{d^{\sigma+1}P_i}{d\mu^{\sigma+1}} - \sigma\left\{(1-\mu^2)^{\frac{\sigma}{2}} - \sigma\mu^2(1-\mu^2)^{\frac{\sigma}{2}-1}\right\}\frac{d^\sigma P_i}{d\mu^\sigma}$$

$$= (1-\mu^2)^{\frac{\sigma}{2}+1}\frac{d^{\sigma+2}P_i}{d\mu^{\sigma+2}} - 2(\sigma+1)\mu(1-\mu^2)^{\frac{\sigma}{2}}\frac{d^{\sigma+1}P_i}{d\mu^{\sigma+1}}$$

$$-\left\{\sigma(\sigma+1)(1-\mu^2)^{\frac{\sigma}{2}} - \sigma^2(1-\mu^2)^{\frac{\sigma}{2}-1}\right\}\frac{d^\sigma P_i}{d\mu^\sigma}.$$

And $\quad i(i+1)\,T_i^{(\sigma)} = i(i+1)(1-\mu^2)^{\frac{\sigma}{2}}\frac{d^\sigma P_i}{d\mu^\sigma};$

$$\therefore\ \frac{d}{d\mu}\left\{(1-\mu^2)\frac{dT_i^{(\sigma)}}{d\mu}\right\} + i(i+1)\,T_i^{(\sigma)}$$

$$= (1-\mu^2)^{\frac{\sigma}{2}+1}\frac{d^{\sigma+2}P_i}{d\mu^{\sigma+2}} - 2(\sigma+1)\mu(1-\mu^2)^{\frac{\sigma}{2}}\frac{d^{\sigma+1}P_i}{d\mu^{\sigma+1}}$$

$$+(i-\sigma)(i+\sigma+1)(1-\mu^2)^{\frac{\sigma}{2}}\frac{d^\sigma P_i}{d\mu^\sigma} + \sigma^2(1-\mu^2)^{\frac{\sigma}{2}-1}\frac{d^\sigma P_i}{d\mu^\sigma}$$

$$= \sigma^2(1-\mu^2)^{\frac{\sigma}{2}-1}\frac{d^\sigma P_i}{d\mu^\sigma} \text{ by (1)}$$

$$= \frac{\sigma^2}{1-\mu^2}\,T_i^{(\sigma)}.$$

Hence the equation above given for M_i is satisfied by $M_i = T_i^{(\sigma)}$, and the equation in Y_i is satisfied by

$$Y_i = CT_i^{(\sigma)}\cos\sigma\phi + C'T_i^{(\sigma)}\sin\sigma\phi.$$

10. In Chap. II. Art. 10 we have established the fundamental property of Zonal Harmonics, that if i and m be two unequal positive integers, $\int_{-1}^{1} P_i P_m d\mu = 0$. This is a particular case of the general theorem that if Y_i, Y_m be two surface harmonics of the degrees i and m respectively,

$$\int_{-1}^{1}\int_0^{2\pi} Y_i Y_m d\mu d\phi = 0.$$

For, let V_i, V_m be the corresponding solid harmonics, so that $V_i = r^i Y_i$, $V_m = r^m Y_m$. Then, by the fundamental property of potential functions, we have at every point at which no attracting matter is situated,

$$\frac{d^2 V_i}{dx^2} + \frac{d^2 V_i}{dy^2} + \frac{d^2 V_i}{dz^2} = 0, \quad \frac{d^2 V_m}{dx^2} + \frac{d^2 V_m}{dy^2} + \frac{d^2 V_m}{dz^2} = 0,$$

and therefore

$$V_i\left(\frac{d^2 V_m}{dx^2} + \frac{d^2 V_m}{dy^2} + \frac{d^2 V_m}{dz^2}\right) - V_m\left(\frac{d^2 V_i}{dx^2} + \frac{d^2 V_i}{dy^2} + \frac{d^2 V_i}{dz^2}\right) = 0,$$

or, in accordance with our notation, $V_i \nabla^2 V_m - V_m \nabla^2 V_i = 0$.

Now, integrate this expression throughout the whole space comprised within a sphere whose centre is the origin and radius a, a being so chosen that this sphere contains no attracting matter. We then have

$$\iiint (V_i \nabla^2 V_m - V_m \nabla^2 V_i)\, dx\, dy\, dz = 0.$$

But also, when the integration extends over all space comprised within any closed surface, we have

$$\iiint (V_i \nabla^2 V_m - V_m \nabla^2 V_i)\, dx dy dz = \iint \left(V_i \frac{dV_m}{dn} - V_m \frac{dV_i}{dn}\right) dS = 0,$$

dS denoting an element of the bounding surface, and $\frac{d}{dn}$ differentiation in the direction of the normal at any point.

Now, in the present case, the bounding surface being a sphere of radius a, and V_i, V_m homogeneous functions of the degrees i, m, respectively,

$$dS = a^2 d\mu d\phi, \quad \frac{dV_i}{dn} = i a^{i-1} Y_i, \quad \frac{dV_m}{dn} = m a^{m-1} Y_m,$$

and, the integration being extended all over the surface of the sphere, the limits of μ are -1 and 1, those of ϕ, 0 and 2π. Hence

$$\iint \left(V_i \frac{dV_m}{dn} - V_m \frac{dV_i}{dn}\right) dS = (m - i)\, a^{i+m+1} \int_{-1}^{1} \int_0^{2\pi} Y_i Y_m d\mu d\phi,$$

whence, *if* m − i *be not* $=0$,

$$\int_{-1}^{1}\int_{0}^{2\pi} Y_i Y_m d\mu d\phi = 0.$$

The value of $\int_{-1}^{1}\int_{0}^{2\pi} Y_i^2 d\mu d\phi$ will be investigated hereafter.

11. We may hence prove that *if a function of* μ *and* ϕ *can be developed in a series of surface harmonics, such development is possible in only one way.*

For suppose, if possible, that there are two such developments, so that

$$F(\mu, \phi) = Y_0 + Y_1 + \ldots + Y_i + \ldots$$

and also

$$F(\mu, \phi) = Y_0' + Y_1' + \ldots + Y_i' + \ldots$$

Then subtracting, we have

$$0 = Y_0 - Y_0' + Y_1 - Y_1' + \ldots + Y_i - Y_i' + \ldots \text{ identically.}$$

Now, each of the expressions $Y_0 - Y_0'$, $Y_1 - Y_1' \ldots Y_i - Y_i'$ being the difference of two surface harmonics of the degree 0, 1, ... i ... is itself a surface harmonic of the degree 0, 1, ... i .... Denote these expressions for shortness by Z_0, $Z_1 \ldots Z_i \ldots$ so that

$$0 = Z_0 + Z_1 + \ldots + Z_i + \ldots \text{ identically.}$$

Then, multiplying by Z_i and integrating all over the surface of the sphere, we have

$$0 = \int_{-1}^{1}\int_{0}^{2\pi} Z_i^2 d\mu d\phi.$$

That is, the sum of an infinite number of essentially positive quantities is $=0$. This can only take place when each of the quantities is separately $=0$. Hence Z_i is identically $=0$, or $Y_i' = Y_i$, and therefore the two developments are identical.

We have not assumed here that such a development is always possible. That it is so, will be shewn hereafter.

12. By referring to the expression for a surface harmonic given in Art. 4, we see that each of the Tesseral and Sectorial Harmonics involves $(1-\mu^2)^{\frac{1}{2}}$, or some power of $(1-\mu^2)^{\frac{1}{2}}$, as a factor, and therefore is equal to 0 when $\mu=\pm 1$. From this it follows that when $\mu=\pm 1$, the value of the Surface Harmonic is independent of ϕ, or that if $Y(\mu, \phi)$ represent a general surface harmonic, $Y(\pm 1, \phi)$ is independent of ϕ, and may therefore be written as $Y(\pm 1)$. Or $Y(1)$ is the value of $Y(\mu, \phi)$ at the pole of the zonal harmonic $P_i(\mu)$, $Y(-1)$ at the other extremity of the axis of $P_i(\mu)$.

We may now prove that

$$\int_0^{2\pi} Y_i d\phi = 2\pi Y_i(1)\, P_i(\mu).$$

For, recurring to the fundamental equation,

$$\frac{d}{d\mu}\left\{(1-\mu^2)\frac{dY_i}{d\mu}\right\}+\frac{1}{1-\mu^2}\frac{d^2 Y_i}{d\phi^2}+i(i+1)Y_i=0.$$

Now, if we integrate this equation with respect to ϕ, between the limits 0 and 2π, we see that, since

$$\int \frac{d^2 Y_i}{d\phi^2}\, d\phi = \frac{dY_i}{d\phi},$$

and the value of Y_i only involves ϕ under the form of cosines or sines of ϕ and its multiples, and therefore the values of $\frac{dY_i}{d\phi}$ are the same at both limits, it follows that

$$\int_0^{2\pi} \frac{d^2 Y_i}{d\phi^2}\, d\phi = 0.$$

Hence

$$\frac{d}{d\mu}\left\{(1-\mu^2)\left(\int_0^{2\pi} Y_i d\phi\right)\right\}+i(i+1)\left(\int_0^{2\pi} Y_i d\phi\right)=0.$$

Hence $\int_0^{2\pi} Y_i d\phi$ is a function of μ which satisfies the fundamental equation for a zonal harmonic, and we therefore have

$$\int_0^{2\pi} Y_i d\phi = CP_i(\mu),$$

C being a constant, as yet unknown.

To determine C, put $\mu=1$, then by the remark just made, Y_i becomes $Y_i(1)$, and is independent of ϕ. Hence, when $\mu=1$, $\int_0^{2\pi} Y_i d\phi = 2\pi Y_i(1)$. Also $P_i(\mu)=1$. We have therefore

$$2\pi Y_i(1) = C,$$

$$\therefore \int_0^{2\pi} Y_i d\phi = 2\pi Y_i(1) P_i(\mu).$$

It follows from this that

$$\int_{-1}^{1}\int_0^{2\pi} P_i Y_i d\mu d\phi = \frac{4\pi}{2i+1} Y_i(1).$$

13. We may now enquire what will be the value of

$$\int_{-1}^{1}\int_0^{2\pi} Y_i Z_i d\mu d\phi,$$

Y_i, Z_i being two general surface harmonics of the degree i. Suppose each to be arranged in a series consisting of the zonal harmonic P_i whose axis is the axis of z, and the system of tesseral and sectorial harmonics deduced from it. Let us represent them as follows:

$$\begin{aligned} Y_i = {} & AP_i \\ & + C_1 T_i^{(1)} \cos\phi + C_2 T_i^{(2)} \cos 2\phi + \ldots + C_\sigma T_i^{(\sigma)} \cos\sigma\phi + \ldots \\ & \qquad + C_i T_i^{(i)} \cos i\phi \\ & + S_1 T_i^{(1)} \sin\phi + S_2 T_i^{(2)} \sin 2\phi + \ldots + S_\sigma T_i^{(\sigma)} \sin\sigma\phi + \ldots \\ & \qquad + S_i T_i^{(i)} \sin i\phi; \end{aligned}$$

$$\begin{aligned} Z_i = {} & aP_i \\ & + c_1 T_i^{(1)} \cos\phi + c_2 T_i^{(2)} \cos 2\phi + \ldots + c_\sigma T_i^{(\sigma)} \cos\sigma\phi + \ldots \\ & \qquad c_\sigma T_i^{(i)} \cos i\phi \\ & + s_1 T_i^{(1)} \sin\phi + s_2 T_i^{(2)} \sin 2\phi + \ldots + s_\sigma T_i^{(\sigma)} \sin\sigma\phi + \ldots \\ & \qquad + s_\sigma T_i^{(i)} \sin i\phi. \end{aligned}$$

Hence the product $Y_i Z_i$ will consist of a series of terms, in which ϕ will enter under the form $\cos\sigma\phi \cos\sigma'\phi$, or $\cos\sigma\phi \sin\sigma'\phi$. This expression when integrated between the limits 0 and 2π vanishes in all cases, except when $\sigma' = \sigma$ and the expression consequently becomes equal to $\cos^2 \sigma\phi$, or $\sin^2 \sigma\phi$. In these cases we know that, σ being any positive integer,

$$\int_0^{2\pi} \cos^2 \sigma\phi \, d\phi = \int_0^{2\pi} \sin^2 \sigma\phi \, d\phi = \pi.$$

Hence the question is reduced to the determination of the value of

$$\int_{-1}^{1} (T_i^{(\sigma)})^2 \, d\mu.$$

Now $T_i^{(\sigma)} = (1-\mu^2)^{\frac{\sigma}{2}} \dfrac{d^\sigma P_i}{d\mu^\sigma}$

$$= \frac{1}{2^i . 1 . 2 . 3 \ldots i} (1-\mu^2)^{\frac{\sigma}{2}} \frac{d^{i+\sigma} (\mu^2-1)^i}{d\mu^{i+\sigma}}.$$

But, by the theorem of Rodrigues, proved in Chap. II. Art. 8, we know that

$$\frac{d^{i+\sigma} (\mu^2-1)^i}{d\mu^{i+\sigma}} = (-1)^\sigma \frac{\lfloor i+\sigma}{\lfloor i-\sigma} (1-\mu^2)^{-\sigma} \frac{d^{i-\sigma} (\mu^2-1)}{d\mu^{i-\sigma}}.$$

Hence $T_i^{(\sigma)}$ may also be expressed under the form

$$(-1)^\sigma \frac{1}{2^i . 1 . 2 . 3 \ldots i} \frac{\lfloor i+\sigma}{\lfloor i-\sigma} (1-\mu^2)^{-\frac{\sigma}{2}} \frac{d^{i-\sigma} (\mu^2-1)^i}{d\mu^{i-\sigma}},$$

whence it follows that

$$(T_i^{(\sigma)})^2 = (-1)^\sigma \left(\frac{1}{2^i . 1 . 2 . 3 \ldots i}\right)^2 \frac{\lfloor i+\sigma}{\lfloor i-\sigma} \frac{d^{i+\sigma} (\mu^2-1)^i}{d\mu^{i+\sigma}} \frac{d^{i-\sigma} (\mu^2-1)^i}{d\mu^{i-\sigma}}.$$

Now, putting $(\mu^2-1)^i = M$ for the moment, and integrating by parts,

$$\int \frac{d^{i+\sigma} M}{d\mu^{i+\sigma}} \frac{d^{i-\sigma} M}{d\mu^{i-\sigma}} d\mu = \frac{d^{i+\sigma-1} M}{d\mu^{i+\sigma-1}} \frac{d^{i-\sigma} M}{d\mu^{i-\sigma}}$$

$$- \int \frac{d^{i+\sigma-1} M}{d\mu^{i+\sigma-1}} \frac{d^{i-\sigma+1} M}{d\mu^{i-\sigma+1}} d\mu.$$

The factor $\frac{d^{i-\sigma}M}{d\mu^{i-\sigma}}$ vanishes at both limits, hence

$$\int_{-1}^{1}\frac{d^{i+\sigma}M}{d\mu^{i+\sigma}}\frac{d^{i-\sigma}M}{d\mu^{i-\sigma}}d\mu=-\int_{-1}^{1}\frac{d^{i+\sigma-1}M}{d\mu^{i+\sigma-1}}\frac{d^{i-\sigma+1}M}{d\mu^{i-\sigma+1}}d\mu$$

$$=(-1)^2\int_{-1}^{1}\frac{d^{i+\sigma-2}M}{d\mu^{i+\sigma-2}}\frac{d^{i-\sigma+2}M}{d\mu^{i-\sigma+2}}d\mu,$$

by a repetition of the same process.

And by repeating this process σ times, we see that

$$\int_{-1}^{1}\frac{d^{i+\sigma}M}{d\mu^{i+\sigma}}\frac{d^{i-\sigma}M}{d\mu^{i-\sigma}}d\mu=(-1)^{\sigma}\int_{-1}^{1}\left(\frac{d^iM}{d\mu^i}\right)^2d\mu$$

$$=(-1)^{\sigma}(2^i.1.2.3\ldots i)^2\int_{-1}^{1}P_i^2\,d\mu$$

$$=(-1)^{\sigma}(2^i.1.2.3\ldots i)^2\frac{2}{2i+1}.$$

Hence $$\int_{-1}^{1}(T_i^{(\sigma)})^2\,d\mu=\frac{\lfloor i+\sigma}{\lfloor i-\sigma}\,\frac{2}{2i+1},$$

and therefore

$$\int_{-1}^{1}\int_{0}^{2\pi}(T_i^{(\sigma)}\cos\sigma\phi)^2\,d\mu d\phi=\int_{-1}^{1}\int_{0}^{2\pi}(T_i^{(\sigma)}\sin\sigma\phi)^2\,d\mu d\phi$$

$$=\frac{\lfloor i+\sigma}{\lfloor i-\sigma}\,\frac{2\pi}{2i+1}.$$

It will be observed that this result does not hold when $\sigma=0$, in which case we have

$$\int_{-1}^{1}\int_{0}^{2\pi}P_i^2\,d\mu d\phi=\frac{4\pi}{2i+1}\text{*}.$$

Hence $\int_{-1}^{1}\int_{0}^{2\pi}Y_iZ_i\,d\mu d\phi$

$$=\frac{4\pi}{2i+1}A_ia_i$$

* In this case $\int_0^{2\pi}\cos^2\sigma\phi\,d\phi=\int_0^{2\pi}\sin^2\sigma\phi\,d\phi=2\pi$.

$$+\frac{2\pi}{2i+1}\left\{\frac{\lfloor i+1}{\lfloor i-1}(C_1c_1+S_1s_1)+\frac{\lfloor i+2}{\lfloor i-2}(C_2c_2+S_2s_2)+\dots\right.$$
$$\left.+\frac{\lfloor i+\sigma}{\lfloor i-\sigma}(C_\sigma c_\sigma+S_\sigma s_\sigma)+\dots+\lfloor 2i\,(C_ic_i+S_is_i)\right\}.$$

14. We have hitherto considered the Zonal Harmonic under its simplest form, that of a "Legendre's Coefficient" in which the axis of z, i.e. the line from which θ is measured, is the axis of the system. We shall now proceed to consider it under the more general form of a "Laplace's Coefficient," in which the axis of the system of zonal harmonics is in any position whatever, and shall shew how this general form may be expressed in terms of $P_i(\mu)$ and of the system of Tesseral and Sectorial Harmonics deduced from it.

Suppose that θ', ϕ' are the angular co-ordinates of the axis of the Zonal Harmonic, i.e. that the angle between this axis and the axis of z is θ', and that the plane containing these two axes is inclined to a fixed plane through the axis of z which we may consider as that of zx, at the angle ϕ'. In accordance with the notation already employed, we shall represent $\cos\theta'$ by μ'.

The rectangular equations of the axis of this system will be

$$\frac{x}{\sin\theta'\cos\phi'}=\frac{y}{\sin\theta'\sin\phi'}=\frac{z}{\cos\theta'}.$$

Hence the Solid Zonal Harmonic of which this is the axis is deduced from the ordinary form of the solid zonal harmonic expressed as a function of z and r by writing, in place of z, $x\sin\theta'\cos\phi'+y\sin\theta'\sin\phi'+z\cos\theta'$.

To deduce the Surface Zonal Harmonic, transform the solid zonal harmonic to polar co-ordinates, by writing $r\sin\theta\cos\phi$ for x, $r\sin\theta\sin\phi$ for y, $r\cos\theta$ for z, and divide by r^i.

The transformation from the special to the general form of surface zonal harmonic may be at once effected, by substituting for μ, or $\cos\theta$, $\cos\theta\cos\theta'+\sin\theta\sin\theta'\cos(\phi-\phi')$.

Now, in order to develope

$$P_i\{\cos\theta\cos\theta'+\sin\theta\sin\theta'\cos(\phi-\phi')\}$$

in the manner already pointed out, assume

$$P_i\{\cos\theta\cos\theta' + \sin\theta\sin\theta'\cos(\phi-\phi')\}$$
$$= AP_i(\mu) + (C^{(1)}\cos\phi + S^{(1)}\sin\phi)\,T_i^{(1)}$$
$$+ (C^{(2)}\cos 2\phi + S^{(2)}\sin 2\phi)\,T_i^{(2)} + \ldots$$
$$+ (C^{(\sigma)}\cos\sigma\phi + S^{(\sigma)}\sin\sigma\phi)\,T_i^{(\sigma)} + \ldots$$
$$+ (C^{(i)}\cos i\phi + S^{(i)}\sin i\phi)\,T_i^{(i)},$$

the letters $A, \ldots C^{(\sigma)}, S^{(\sigma)}\ldots$ denoting functions of μ' and ϕ', to be determined.

To determine $C^{(\sigma)}$, multiply both sides of this equation by $\cos\sigma\phi\, T_i^{(\sigma)}$ and integrate all over the surface of the sphere, i.e. between the limits -1 and 1 of μ, and 0 and 2π of ϕ. We then get

$$\int_{-1}^{1}\int_{0}^{2\pi} P_i\{\cos\theta\cos\theta' + \sin\theta\sin\theta'\cos(\phi-\phi')\}\cos\sigma\phi\, T_i^{(\sigma)}\, d\mu d\phi$$

$$= C^{(\sigma)}\int_{-1}^{1}\int_{0}^{2\pi}(\cos\sigma\phi\, T_i^{(\sigma)})^2\, d\mu d\phi$$

$$= \frac{\underline{|i+\sigma}}{\underline{|i-\sigma}}\,\frac{2\pi}{2i+1}\,C^{(\sigma)}.$$

It remains to find the value of the left-hand member of this equation.

Now $\cos\sigma\phi\, T_i^{(\sigma)}$ is a surface harmonic of the degree i, and therefore a function of the kind denoted by Y_i in Art. 12.

And we have shewn, in that Article, that

$$\int_{-1}^{1}\int_{0}^{2\pi} P_i(\mu)\, Y_i\, d\mu d\phi = \frac{4\pi}{2i+1}\, Y_i(1),$$

that is, that *if any surface harmonic of the degree* i *be multiplied by the zonal harmonic of the same degree, and the product integrated all over the surface of the sphere, the integral is equal to* $\frac{4\pi}{2i+1}$ *into the value which the surface harmonic assumes at the pole of the zonal harmonic.*

Hence

$$\int_{-1}^{1}\int_{0}^{2\pi} P_i\{\cos\theta\cos\theta' + \sin\theta\sin\theta'\cos(\phi-\phi')\} Y_i(\mu,\phi)\,d\mu\,d\phi$$
$$= \frac{4\pi}{2i+1} Y_i(\mu',\phi'),$$

and therefore

$$\int_{-1}^{1}\int_{0}^{2\pi} P_i\{\cos\theta\cos\theta' + \sin\theta\sin\theta'\cos(\phi-\phi')\} \cos\sigma\phi T_i^{(\sigma)}\,d\mu\,d\phi$$
$$= \frac{4\pi}{2i+1}\cos\sigma\phi'\, T_i^{(\sigma)}\,\mu'.$$

Hence

$$\frac{4\pi}{2i+1}\cos\sigma\phi'\, T_i^{(\sigma)}(\mu') = \frac{\lfloor i+\sigma}{\lfloor i-\sigma}\,\frac{2\pi}{2i+1}\,C^{(\sigma)},$$

$$\text{or } C^{(\sigma)} = 2\frac{\lfloor i-\sigma}{\lfloor i+\sigma}\cos\sigma\phi'\, T_i^{(\sigma)}(\mu').$$

Similarly $$S^{(\sigma)} = 2\frac{\lfloor i-\sigma}{\lfloor i+\sigma}\sin\sigma\phi'\, T_i^{(\sigma)}(\mu').$$

And to determine A, we have

$$\int_{-1}^{1}\int_{0}^{2\pi} P_i\{\cos\theta\cos\theta' + \sin\theta\sin\theta'\cos(\phi-\phi')\} P_i(\mu)\,d\mu\,d\phi$$
$$= A\int_{-1}^{1}\int_{0}^{2\pi}\{P_i(\mu)\}^2\,d\mu\,d\phi;$$

$$\therefore \frac{4\pi}{2i+1}P_i(\mu') = A\frac{4\pi}{2i+1},$$
$$\text{or } A = P_i(\mu').$$

Hence, $P_i\{\cos\theta\cos\theta' + \sin\theta\sin\theta'\cos(\phi-\phi')\}$

$$= P_i(\mu')P_i(\mu) + 2\frac{\lfloor i-1}{\lfloor i+1}\cos(\phi-\phi')\,T_i^{(1)}(\mu')\,T_i^{(1)}(\mu)$$
$$+ 2\frac{\lfloor i-2}{\lfloor i+2}\cos 2(\phi-\phi')\,T_i^{(2)}(\mu')\,T_i^{(2)}(\mu) + \ldots$$

$$+2\frac{\lfloor i-\sigma}{\lfloor i+\sigma}\cos\sigma(\phi-\phi')\,T_i^{(\sigma)}(\mu')\,T_i^{(\sigma)}(\mu)+\dots$$

$$+2\frac{1}{\lfloor 2i}\cos i(\phi-\phi')\,T_i^{(i)}(\mu')\,T_i^{(i)}(\mu).$$

15. We have already seen (Chap. II. Art. 20) how any rational integral function of μ can be expressed by a finite series of zonal harmonics. We shall now shew how any rational integral function of $\cos\theta$, $\sin\theta\cos\phi$, $\sin\theta\sin\phi$, can be expressed by a finite series of zonal, tesseral, and sectorial harmonics.

For any power of $\cos\phi$ or $\sin\phi$, or any product of such powers, may be expressed as the sum of a series of terms of the form $\cos\sigma\phi$, or $\sin\sigma\phi$, the greatest value of σ being the sum of the indices of $\cos\phi$ and $\sin\phi$, and the other values diminishing by 2 in each successive term. Hence any rational integral function of $\cos\theta$, $\sin\theta\cos\phi$, $\sin\theta\sin\phi$, will consist of a series of terms of the form

$$\cos^m\theta\sin^n\theta\cos\sigma\phi \text{ or } \cos^m\theta\sin^n\theta\sin\sigma\phi,$$

where n is not less than σ.

If n be greater than σ, $n-\sigma$ must be an even integer. Let $n-\sigma=2s$, then writing $\sin^n\theta$ under the form $(1-\cos^2\theta)^s\sin^\sigma\theta$, we reduce $\cos^m\theta\sin^n\theta\cos\sigma\phi$ to the sum of a series of terms of the form $\cos^p\theta\sin^\sigma\theta\cos\sigma\phi$, or, writing $\cos\theta=\mu$, of the form $\mu^p(1-\mu^2)^{\frac{\sigma}{2}}\cos\sigma\phi$.

Similarly $\cos^m\theta\sin^n\theta\sin\sigma\phi$ is reduced to a series of terms of the form $\mu^p(1-\mu^2)^{\frac{\sigma}{2}}\sin\sigma\phi$.

Now $$\mu^p=\frac{1}{(p+\sigma)(p+\sigma-1)\dots(p+1)}\frac{d^\sigma}{d\mu^\sigma}\mu^{p+\sigma},$$

and $\mu^{p+\sigma}$ can be developed in a series of terms of the form of multiples of $P_{p+\sigma}$, $P_{p+\sigma-2}\dots$. (Chap. II. Art. 17.)

Hence μ^p can be expressed in a series of the form

$$\frac{d^\sigma}{d\mu^\sigma}(A_0P_{p+\sigma}+A_2P_{p+\sigma-2}+\dots),$$

A_0, A_1 representing known numerical constants, and therefore $\mu^p (1-\mu^2)^{\frac{\sigma}{2}}$ assumes the form

$$(A_0\, T_{p+\sigma}^{(\sigma)} + A_2\, T_{p+\sigma-2}^{(\sigma)} + \ldots),$$

consequently multiplying these series by $\cos \sigma\phi$ or $\sin \sigma\phi$, we obtain the developments of

$$\mu^p (1-\mu^2)^{\frac{\sigma}{2}} \cos \sigma\phi \text{ and } \mu^p (1-\mu^2)^{\frac{\sigma}{2}} \sin \sigma\phi$$

in series of tesseral harmonics.

16. We will give two illustrations of this transformation.

First, suppose it is required to express $\cos^2\theta \sin^2\theta \sin\phi \cos\phi$ in a series of Spherical Harmonics.

Here we have $\sin\phi\cos\phi = \frac{1}{2}\sin 2\phi.$

Hence $\cos^2\theta \sin^2\theta \sin\phi\cos\phi = \frac{1}{2}\cos^2\theta \sin^2\theta \sin 2\phi.$

Comparing this with $\cos^m\theta \sin^n\theta \sin\sigma\phi$, we see that n is not greater than σ.

Hence $\cos^2\theta \sin^2\theta \sin\phi\cos\phi = \frac{1}{2}\mu^2(1-\mu^2)\sin 2\phi.$

And
$$\mu^2 = \frac{1}{4\,.\,3}\frac{d^2}{d\mu^2}\mu^4,$$

and
$$\mu^4 = \frac{8}{35}P_4 + \frac{4}{7}P_2 + \frac{1}{5}P_0,$$

$$\therefore\ \mu^2 = \frac{1}{12}\left(\frac{8}{35}\frac{d^2P_4}{d\mu^2} + \frac{4}{7}\frac{d^2P_2}{d\mu^2}\right)$$

$$= \frac{2}{105}\frac{d^2P_4}{d\mu^2} + \frac{1}{21}\frac{d^2P_2}{d\mu^2},$$

$\therefore\ \cos^2\theta\sin^2\theta\sin\phi\cos\phi$

$$= \frac{1}{2}\left\{\frac{2}{105}\frac{d^2P_4}{d\mu^2}(1-\mu^2)\sin 2\phi + \frac{1}{21}\frac{d^2P_2}{d\mu^2}(1-\mu^2)\sin 2\phi\right\}$$

$$= \left\{\frac{1}{105}T_4^{(2)} + \frac{1}{42}T_2^{(2)}\right\}\sin 2\phi.$$

Next, let it be required to transform $\cos^3\theta \sin^3\theta \sin\phi \cos^2\phi$ into a series of Spherical Harmonics.

Here $\sin\phi\cos^2\phi = \frac{1}{2}\sin 2\phi \cos\phi = \frac{1}{4}(\sin 3\phi + \sin\phi)$.

Now $\cos^3\theta \sin^3\theta \sin 3\phi = \mu^3 (1-\mu^2)^{\frac{3}{2}} \sin 3\phi$

$$= \frac{1}{6\,.\,5\,.\,4} \frac{d^3}{d\mu^3}\, \mu^6\,.\,(1-\mu^2)^{\frac{3}{2}} \sin 3\phi.$$

Also $\cos^3\theta \sin^3\theta \sin\phi = \mu^3 (1-\mu^2)(1-\mu^2)^{\frac{1}{2}} \sin\phi$

$$= (\mu^3 - \mu^5)(1-\mu^2)^{\frac{1}{2}} \sin\phi$$

$$= \left(\frac{1}{4}\frac{d}{d\mu}\mu^4 - \frac{1}{6}\frac{d}{d\mu}\mu^6\right)(1-\mu^2)^{\frac{1}{2}} \sin\phi.$$

Also (Chap. II. Art. 17)

$$\mu^4 = \frac{8}{35}P_4 + \frac{4}{7}P_2 + \frac{1}{5}P_0,$$

$$\mu^6 = \frac{16}{231}P_6 + \frac{24}{77}P_4 + \frac{10}{21}P_2 + \frac{1}{7}P_0.$$

Hence $\cos^3\theta \sin^3\theta \sin 3\phi$

$$= \frac{1}{120}\left(\frac{16}{231}\frac{d^3P_6}{d\mu^3} + \frac{24}{77}\frac{d^3P_4}{d\mu^3}\right)(1-\mu^2)^{\frac{3}{2}} \sin 3\phi$$

$$= \left\{\frac{2}{3465}T_6^{(3)} + \frac{1}{385}T_4^{(3)}\right\} \sin 3\phi.$$

And $\cos^3\theta \sin^3\theta \sin\phi = -\left(\frac{8}{693}\frac{dP_6}{d\mu} + \frac{4}{77}\frac{dP_4}{d\mu} + \frac{5}{63}\frac{dP_2}{d\mu}\right.$

$$\left. - \frac{2}{35}\frac{dP_4}{d\mu} - \frac{1}{7}\frac{dP_2}{d\mu}\right)(1-\mu^2)^{\frac{1}{2}} \sin\phi$$

$$= -\left(\frac{8}{693}\frac{dP_6}{d\mu} - \frac{2}{385}\frac{dP_4}{d\mu} - \frac{4}{63}\frac{dP_2}{d\mu}\right)(1-\mu^2)^{\frac{1}{2}} \sin\phi$$

$$= -\left(\frac{8}{693}T_6^{(1)} - \frac{2}{385}T_4^{(1)} - \frac{4}{63}T_2^{(1)}\right) \sin\phi;$$

$$\therefore\ \cos^3\theta \sin^3\theta \sin\phi \cos^2\phi = \left\{\frac{1}{6930}T_6^{(3)} + \frac{1}{1540}T_4^{(3)}\right\} \sin 3\phi$$

$$- \left\{\frac{2}{693}T_6^{(1)} - \frac{1}{770}T_4^{(1)} - \frac{1}{63}T_2^{(1)}\right\} \sin\phi.$$

17. The process above investigated is probably the most convenient one when the object is to transform any finite algebraical function of $\cos\theta$, $\sin\theta\cos\phi$, and $\sin\theta\sin\phi$, into a series of spherical harmonics. For general forms of a function of μ and ϕ, however, this method is inapplicable, and we proceed to investigate a process which will apply universally, even if the function to be transformed be discontinuous.

We must first discuss the following problem.

To determine the potential of a spherical shell whose surface density is $F(\mu, \phi)$, F denoting any function whatever of finite magnitude, at an external or internal point.

Let c be the radius of the sphere, r' the distance of the point from its centre, θ', ϕ' its angular co-ordinates, V the potential. Then μ being equal to $\cos\theta$

$$V=\int_{-1}^{1}\int_{0}^{2\pi}\frac{F(\mu, \phi)\, c^2\, d\mu\, d\phi}{[r'^2-2cr'\{\cos\theta\cos\theta'+\sin\theta\sin\theta'\cos(\phi-\phi')\}+c^2]^{\frac{1}{2}}}.$$

The denominator, when expanded in a series of general zonal harmonics, or Laplace's coefficients, becomes

$$\frac{1}{c}\left\{1+P_1(\mu, \phi)\frac{r'}{c}+P_2(\mu, \phi)\frac{r'^2}{c^2}+\ldots+P_i(\mu, \phi)\frac{r'^i}{c^i}+\ldots\right\},$$

$$\frac{1}{r'}\left\{1+P_1(\mu, \phi)\frac{c}{r'}+P_2(\mu, \phi)\frac{c^2}{r'^2}+\ldots+P_i(\mu, \phi)\frac{c^i}{r'^i}+\ldots\right\},$$

for an internal and an external point respectively, $P_i(\mu, \phi)$ being written for

$$P_i\{\cos\theta\cos\theta'+\sin\theta\sin\theta'\cos(\phi-\phi')\}.$$

Hence, V_1 denoting the potential at an internal, V_2 at an external, point,

$$V_1=c\left\{\int_{-1}^{1}\int_{0}^{2\pi}F(\mu, \phi)\, d\mu\, d\phi+\frac{r'}{c}\int_{-1}^{1}\int_{0}^{2\pi}P_1(\mu, \phi)\, F(\mu, \phi)\, d\mu\, d\phi\right.$$

$$\left.+\ldots+\frac{r'^i}{c^i}\int_{-1}^{1}\int_{0}^{2\pi}P_i(\mu, \phi)\, F(\mu, \phi)\, d\mu\, d\phi+\ldots\right\},$$

$$V_2 = \frac{c^2}{r'}\left\{\int_{-1}^{1}\int_0^{2\pi} F(\mu, \phi)\, d\mu\, d\phi + \frac{c}{r'}\int_{-1}^{1}\int_0^{2\pi} P_1(\mu, \phi)\, F(\mu, \phi)\, d\mu\, d\phi\right.$$
$$\left. + \ldots + \frac{c^i}{r'^i}\int_{-1}^{1}\int_0^{2\pi} P_i(\mu, \phi)\, F(\mu, \phi)\, d\mu\, d\phi + \ldots\right\}.$$

It will be observed that the expression $P_i(\mu, \phi)$ involves μ and μ' symmetrically, and also ϕ and ϕ'. Hence it satisfies the equation

$$\frac{d}{d\mu'}\left\{(1-\mu'^2)\frac{dP'}{d\mu'}\right\} + \frac{1}{1-\mu'^2}\frac{d^2P'}{d\mu'^2} + i(i+1)P_i = 0.$$

And, since μ and ϕ are independent of μ' and ϕ', this differential equation will continue to be satisfied after P_i has been multiplied by any function of μ and ϕ, and integrated with respect to μ and ϕ. That is, every expression of the form

$$\int_{-1}^{1}\int_0^{2\pi} P_i(\mu, \phi)\, F(\mu, \phi)\, d\mu\, d\phi$$

is a Spherical Surface Harmonic, or "Laplace's Function" with respect to μ' and ϕ' of the degree i. And the several terms of the developments of V_1 are solid harmonics of the degree $0, 1, 2 \ldots i \ldots$ while those of V_2 are the corresponding functions of the degrees $-1, -2, -3 \ldots -(i+1), \ldots$ And these are the expressions for the potential at a point (r', μ', ϕ') of the distribution of density $F(\mu', \phi')$ at a point (c, μ', ϕ').

Now, the expressions for the potentials, both external and internal, given in the last Article, are precisely the same as those for the distribution of matter whose surface density is

$$\frac{1}{4\pi}\left\{\int_{-1}^{1}\int_0^{2\pi} F(\mu, \phi)\, d\mu\, d\phi + 3\int_{-1}^{1}\int_0^{2\pi} P_1(\mu, \phi)\, F(\mu, \phi)\, d\mu\, d\phi + \ldots\right.$$
$$\left. + (2i+1)\int_{-1}^{1}\int_0^{2\pi} P_i(\mu, \phi)\, F(\mu, \phi)\, d\mu\, d\phi + \ldots\right\},$$

or, as it may now be better expressed,

$$\frac{1}{4\pi}\left[\int_{-1}^{1}\int_0^{2\pi} F(\mu, \phi)\, d\mu\, d\phi\right.$$

$$+3\int_{-1}^{1}\int_{0}^{2\pi} P_1\{\cos\theta\cos\theta' + \sin\theta\sin\theta'\cos(\phi-\phi') F(\mu,\phi)\,d\mu d\phi$$

$$+\ldots$$

$$+(2i+1)\int_{-1}^{1}\int_{0}^{2\pi} P_i\{\cos\theta\cos\theta' + \sin\theta\sin\theta'\cos(\phi-\phi')\}F(\mu,\phi)\,d\mu d\phi + \ldots\Big].$$

And, since there is only one distribution of density which will produce a given potential at every point both external and internal, it follows that this series must be identical with $F(\mu', \phi')$. We have thus, therefore, investigated the development of $F(\mu', \phi')$ in a series of spherical surface harmonics*.

The only limitation on the generality of the function $F(\mu', \phi')$ is that it should not become infinite for any pair of values comprised between the limits -1 and 1 of μ, and 0 and 2π of ϕ.

18. Ex. To express $\cos 2\phi'$ in a series of spherical harmonics.

For this purpose, it is necessary to determine the value of

$$(2i+1)\int_{-1}^{1}\int_{0}^{2\pi} P_i\{\cos\theta\cos\theta' + \sin\theta\sin\theta'\cos(\phi-\phi')\}\cos 2\phi\,d\mu d\phi.$$

Now $\qquad P_i\{\cos\theta\cos\theta' + \sin\theta\sin\theta'\cos(\phi-\phi')\}$

$$= P_i(\cos\theta)\,P_i(\cos\theta')$$

$$+\frac{2}{i(i+1)}\sin\theta\,\frac{dP_i(\cos\theta)}{d\mu}\sin\theta'\,\frac{dP_i(\cos\theta')}{d\mu'}\cos(\phi-\phi')$$

$$+\frac{2}{(i-1)\,i\,(i+1)(i+2)}\sin^2\theta$$

$$\frac{d^2P_i(\cos\theta)}{d\mu^2}\sin^2\theta'\,\frac{d^2P_i(\cos\theta')}{d\mu'^2}\cos 2(\phi-\phi') + \ldots$$

Now $\qquad \displaystyle\int_0^{2\pi}\cos\sigma(\phi-\phi')\cos 2\phi\,d\phi = 0,$

for all values of σ except 2.

* In connection with the subject of this Article, see a paper by Mr G. H. Darwin in the *Messenger of Mathematics* for March, 1877.

And $\int_0^{2\pi} \cos 2(\phi - \phi') \cos 2\phi \, d\phi = \pi \cos 2\phi'$.

Also

$$\int \sin^2\theta \frac{d^2 P_i}{d\mu^2} d\mu = -\frac{1}{2^i . 1.2.3 \ldots i} \int (\mu^2 - 1) \frac{d^{i+2} (\mu^2-1)^i}{d\mu^{i+2}} d\mu.$$

And

$$\int (\mu^2 - 1) \frac{d^{i+2} (\mu^2-1)^i}{d\mu^{i+2}} d\mu = (\mu^2 - 1) \frac{d^{i+1} (\mu^2-1)^i}{d\mu^{i+1}} - 2 \int \mu \frac{d^{i+1} (\mu^2-1)^i}{d\mu^{i+1}} d\mu,$$

$$\int \mu \frac{d^{i+1} (\mu^2-1)^i}{d\mu^{i+1}} d\mu = \mu \frac{d^i (\mu^2-1)^i}{d\mu^i} - \frac{d^{i-1} (\mu^2-1)^i}{d\mu^{i-1}}$$

$$= 2^i . 1.2.3 \ldots i \mu P_i - \frac{d^{i-1} (\mu^2-1)^i}{d\mu^{i-1}}.$$

Now when $\mu = 1$,

$$(\mu^2 - 1) \frac{d^{i+1} (\mu^2-1)^i}{d\mu^{i+1}} = 0, \quad \mu P_i = 1, \quad \frac{d^{i-1} (\mu^2-1)^i}{d\mu^{i-1}} = 0.$$

And when $\mu = -1$,

$$(\mu^2 - 1) \frac{d^{i+1} (\mu^2-1)^i}{d\mu^{i+1}} = 0, \quad \mu P_i = (-1)^{i+1}, \quad \frac{d^{i-1} (\mu^2-1)^i}{d\mu^{i-1}} = 0.$$

Hence

$$\int_{-1}^{1} \sin^2\theta \frac{d^2 P_i}{d\mu^2} d\mu = \frac{2}{2^i . 1.2.3 \ldots i} 2^i . 1.2.3 \ldots i \{1 - (-1)^{i+1}\}$$

$$= 4 \text{ or } 0, \text{ as } i \text{ is even or odd};$$

$$\therefore \int_{-1}^{1} \int_0^{2\pi} \sin^2\theta \frac{d^2 P_i(\cos\theta)}{d\mu^2} \cos 2(\phi - \phi') \cos 2\phi \, d\mu \, d\phi$$

$$= 4\pi \cos 2\phi' \text{ or } 0, \text{ as } i \text{ is even or odd};$$

$\therefore \cos 2\phi'$

$$= \frac{1}{4\pi} \left\{ 5 \frac{2}{1.2.3.4} 4 \sin^2\theta' \frac{d^2 P_2(\cos\theta')}{d\mu'^2} \pi \cos 2\phi' \right.$$

$$+ \; 9 \frac{2}{3.4.5.6} 4 \sin^4\theta' \frac{d^2 P_4(\cos\theta')}{d\mu'^2} \pi \cos 2\phi'$$

$$+ 13 \frac{2}{5.6.7.8} 4 \sin^6 \theta' \frac{d^2 P_6(\cos\theta')}{d\mu'^2} \pi \cos 2\phi'$$

$$+ \ldots\ldots\Big\}$$

$$= 2\cos 2\phi' \left(\frac{5.T_2^{(2)}}{1.2.3.4} + \frac{9.T_4^{(2)}}{3.4.5.6} + \frac{13.T_6^{(2)}}{5.6.7.8} + \ldots\right).$$

Hence the potential of a spherical shell, of radius c and surface density $\cos 2\phi'$, will be

$$8\pi \cos 2\phi' \left(\frac{T_2^{(2)}}{1.2.3.4}\frac{r'^2}{c^2} + \frac{T_4^{(2)}}{3.4.5.6}\frac{r'^4}{c^4} + \frac{T_6^{(2)}}{5.6.7.8}\frac{r'^6}{c^6} + \ldots\right),$$

and

$$8\pi \cos 2\phi' \left(\frac{T_2^{(2)}}{1.2.3.4}\frac{c^3}{r'^3} + \frac{T_4^{(2)}}{3.4.5.6}\frac{c^5}{r'^5} + \frac{T_6^{(2)}}{5.6.7.8}\frac{c^7}{r'^7} + \ldots\right),$$

at an internal and external point respectively.

19. We will now explain the application of Spherical Harmonics to the determination of the potential of a homogeneous solid, nearly spherical in form. The following investigation is taken from the *Mécanique Céleste*, Liv. III. Chap. II.

Let r be the radius vector of such a solid, and let

$$r = a + \alpha (a_1 Y_1 + a_2 Y_2 + \ldots + a_i Y_i + \ldots),$$

α being a small quantity, whose square and higher powers may be neglected, $a_1, a_2, \ldots a_i \ldots$ lines of arbitrary length, and $Y_1, Y_2, \ldots Y_i \ldots$ surface harmonics of the order $1, 2, \ldots i \ldots$ respectively.

The volume of the solid will be $\frac{4}{3}\pi a^3$.

For it is equal to

$$\int_0^r \int_{-1}^1 \int_0^{2\pi} r^2\, dr\, d\mu\, d\phi$$

$$= \frac{1}{3}\int_{-1}^1 \int_0^{2\pi} \{a^3 + 3a^2\alpha (a_1 Y_1 + a_2 Y_2 + \ldots + a_i Y_i + \ldots)\}\, d\mu\, d\phi$$

$$= \frac{4}{3}\pi a^3, \text{ since } \int_{-1}^1 \int_0^{2\pi} Y_i\, d\mu\, d\phi = 0,$$

for all values of i.

Again, if the centre of gravity of the solid be taken as origin, $a_1 = 0$.

For if $\bar{z}$ be the distance of the centre of gravity from the plane of xy,

$$\frac{4}{3}\pi a^3 \bar{z} = \int_0^r \int_{-1}^1 \int_0^{2\pi} r^3 \mu \, dr \, d\mu \, d\phi$$

$$= \frac{1}{4} \int_{-1}^1 \int_0^{2\pi} a^4 + 4a^3 \alpha \, (a_1 Y_1 + a_2 Y_2 + \ldots + a_i Y_i + \ldots) \, \mu \, d\mu \, d\phi$$

$$= 4a^3 \alpha \, a_1 \int_{-1}^1 \int_0^{2\pi} \mu Y_1 \, d\mu \, d\phi.$$

Similarly

$$\frac{4}{3}\pi a^3 \bar{x} = 4a^3 \alpha \,.\, a_1 \int_{-1}^1 \int_0^{2\pi} (1 - \mu^2)^{\frac{1}{2}} \cos\phi \, Y_1 \, d\mu \, d\phi,$$

$$\frac{4}{3}\pi a^3 \bar{y} = 4a^3 \alpha \,.\, a_1 \int_{-1}^1 \int_0^{2\pi} (1 - \mu^2)^{\frac{1}{2}} \sin\phi \, Y_1 \, d\mu \, d\phi.$$

Now Y_1 is an expression of the form

$$A\mu + B(1 - \mu^2)^{\frac{1}{2}} \cos\phi + C(1 - \mu^2)^{\frac{1}{2}} \sin\phi,$$

and therefore all the expressions $\bar{x}$, $\bar{y}$, $\bar{z}$ cannot be equal to 0, unless $a_1 = 0$.

We may therefore, taking the centre of gravity as origin, write

$$r = a + \alpha \, (a_2 Y_2 + \ldots + a_i Y_i + \ldots),$$

as the equation of the bounding surface of the solid.

Now this solid may be considered as made up of a homogeneous sphere, radius a, and of a shell, whose thickness is

$$\alpha \, (a_2 Y_2 + \ldots + a_i Y_i + \ldots).$$

The potential of this shell, at least at points whose least distance from it is considerable compared with its thickness, will be the same as that of a shell whose thickness is αa, and density

$$\rho_0 \left(\frac{a_2}{a} Y_2 + \ldots + \frac{a_i}{a} Y_i + \ldots \right),$$

ρ_0 being the density of the solid. Therefore the potential, for any external point, distant R from the centre, will be

$$4\pi\rho_0 \frac{a^3}{3R} + 4\pi\rho_0 \alpha a^2 \left(\frac{a_2 Y_2}{5} \frac{a^2}{R^3} + \dots + \frac{a_i Y_i}{2i+1} \frac{a^i}{R^{i+1}} + \dots\right).$$

The potential at any internal point, distant R from the centre, will be made up of the two portions

$$\frac{4}{3}\pi\rho_0 R^2 + 2\pi\rho_0 (a^2 - R^2) \text{ or } 2\pi\rho \left(a^2 - \frac{R^2}{3}\right)$$

for the homogeneous sphere,

$$4\pi\rho_0 \alpha a^2 \left(\frac{a_2 Y_2}{5} \frac{R^2}{a^3} + \dots + \frac{a_i Y_i}{2i+1} \frac{R^i}{a^{i+1}} + \dots\right)$$

for the shell, and will therefore be equal to

$$2\pi\rho_0 \left(a^2 - \frac{R^2}{3}\right) + 4\pi\rho_0 \alpha a^2 \left(\frac{a_2 Y_2}{5} \frac{R^2}{a^3} + \dots + \frac{a_i Y_i}{2i+1} \frac{R^i}{a^{i+1}} + \dots\right).$$

20. If the solid, instead of being homogeneous, be made up of strata of different densities, the strata being concentric, and similar to the bounding surface of the solid, we may deduce an expression for its potential as follows. Let $\frac{c}{a} r$ be the radius vector of any stratum, ρ its density, r having the same value as in the last Article, and ρ being a function of c only. Then, δc being the mean thickness of the stratum, that is the difference between the values of c for its inner and outer surfaces, the potential of the stratum at an external point will be

$$\frac{4\pi\rho c^2 \delta c}{R} + 4\pi\rho\alpha \frac{c^2 \delta c}{a} \left(\frac{a_2 Y_2}{5} \frac{c^2}{R^2} + \frac{a_3 Y_3}{7} \frac{c^3}{R^3} + \dots \right.$$
$$\left. + \frac{a_i Y_i}{2i+1} \frac{c^i}{R^i} + \dots\right) \dots\dots\dots (1).$$

To obtain the potential of the whole solid at an external point we must integrate this expression with respect to c, between the limits 0 and a, remembering that ρ is a function of c.

Again, the potential of the stratum, above considered, at an internal point will be

$$4\pi\rho c\delta c + 4\pi\rho\alpha \frac{c^2\delta c}{a}\left(\frac{a_2 Y_2}{5}\frac{R^2}{c^2} + \frac{a_3 Y_3}{7}\frac{R^3}{c^3} + \dots \right.$$

$$\left. + \frac{a_i Y_i}{2i+1}\frac{R^i}{c^i} + \dots\right) \dots\dots\dots (2).$$

To obtain the potential of the whole solid at an internal point we must integrate the expression (1) with respect to c between the limits 0 and R, and the expression (2) with respect to c between the limits R and a, remembering in both cases that ρ is a function of c, and add the results together.

The (attr) potential for a pt. on the surface of a body nearly a sphere is

$$V = ga + 3g\sum_{i=2}\frac{S_i}{2i+1}$$

where $S_i = \alpha a_i Y_i$

[Vide application of this [illegible] in Craig's [illegible] [illegible] problem solved by Sir W. Thomson.]

CHAPTER V.

SPHERICAL HARMONICS OF THE SECOND KIND.

1. We have already seen (Chap. II. Art. 2) that the differential equation of which P_i is one solution, being of the second order, admits of another solution, viz.

$$CP_i \int \frac{d\mu}{P_i^2 (1-\mu^2)}.$$

Now if μ between the limits of integration be equal to ± 1, or to any roots of the equation $P_i = 0$ (all of which roots lie between 1 and -1), the expression under the integral sign becomes infinite between the limits of integration. We can therefore only assign an intelligible meaning to this integral, by supposing μ to be always between 1 and ∞, or between -1 and $-\infty$. We will adopt the former supposition, and if we then put $C = -1$, the expression $\dfrac{C}{P_i^2 (1-\mu^2)}$ $\left(\text{i.e. } \dfrac{1}{P_i^2 (\mu^2 - 1)}\right)$ will be always positive. We may therefore define the expression

$$P \int_\mu^\infty \frac{d\mu}{P_i^2 (\mu^2 - 1)},$$

as the zonal harmonic of the second kind, which we shall denote by Q_i, or $Q_i(\mu)$, when it is necessary to specify the variables of which it is a function.

It will be observed that, if μ be greater than 1, P_i is always positive. Hence, on the same supposition, Q_i is always positive.

We see that $Q_0 = \displaystyle\int_\mu^\infty \frac{d\mu}{\mu^2 - 1} = \frac{1}{2} \log \frac{\mu + 1}{\mu - 1}$,

$$Q_1 = \mu \int_\mu^\infty \frac{d\mu}{\mu^2(\mu^2-1)}$$
$$= \mu \int_\mu^\infty \left(\frac{1}{\mu^2-1} - \frac{1}{\mu^2}\right) d\mu$$
$$= \frac{1}{2}\mu \log \frac{\mu+1}{\mu-1} - 1.$$

And, in a similar manner, the values of Q_2, Q_3,... may be calculated.

2. But there is another manner of arriving at these functions, which will enable us to express them, when the variable is greater than unity, in a converging series, without the necessity of integration.

This we shall do in the following manner.

Let $U = \dfrac{1}{\nu - \mu}$, ν being not less, and μ not greater, than unity.

Then
$$\frac{dU}{d\nu} = -\frac{1}{(\nu-\mu)^2}, \qquad \frac{dU}{d\mu} = \frac{1}{(\nu-\mu)^2},$$
$$(1-\nu^2)\frac{dU}{d\nu} = \frac{\nu^2-1}{(\nu-\mu)^2}, \quad (1-\mu^2)\frac{dU}{d\mu} = \frac{1-\mu^2}{(\nu-\mu)^2},$$
$$\frac{d}{d\nu}\left\{(1-\nu^2)\frac{dU}{d\nu}\right\} = \frac{\nu^2-1}{(\nu-\mu^2)}\left(\frac{2\nu}{\nu^2-1} - \frac{2}{\nu-\mu}\right) = 2\frac{1-\mu\nu}{(\nu-\mu)^3},$$
$$\frac{d}{d\mu}\left\{(1-\mu^2)\frac{dU}{d\mu}\right\} = \frac{1-\mu^2}{(\nu-\mu)^2}\left(-\frac{2\mu}{1-\mu^2} + \frac{2}{\nu-\mu}\right) = 2\frac{1-\mu\nu}{(\nu-\mu)^3};$$
$$\therefore \frac{d}{d\nu}\left\{(1-\nu^2)\frac{dU}{d\nu}\right\} = \frac{d}{d\mu}\left\{(1-\mu^2)\frac{dU}{d\mu}\right\}.$$

Now, let $\dfrac{1}{\nu-\mu}$ be expanded in a series of zonal harmonics $P_0(\mu)$, $P_1(\mu)\ldots P_i(\mu)$, so that

$$U = \frac{1}{\nu-\mu} = \phi_0(\nu)P_0(\mu) + \phi_1(\mu)P_1(\mu) + \ldots + \phi_i(\nu)P_i(\mu) + \ldots$$

Then $\dfrac{d}{d\mu}\left\{(1-\mu^2)\dfrac{dU}{d\mu}\right\} = \ldots - i(i+1)\phi_i(\nu)P_i(\mu) + \ldots$

by the definition of $P(\mu)$.

And also $\frac{d}{d\nu}\left\{(1-\nu^2)\frac{dU}{d\nu}\right\} = \ldots + \frac{d}{d\nu}\left\{(1-\nu^2)\frac{d\phi_i(\nu)}{d\nu}\right\} P_i(\mu) + \ldots$

And these two expressions are equal. Hence, equating the coefficients of $P_i(\mu)$,

$$\frac{d}{d\nu}\left\{(1-\nu^2)\frac{d\phi_i(\nu)}{d\nu}\right\} = -i(i+1)\,\phi_i(\nu).$$

Hence $\phi_i(\nu)$ satisfies the same differential equation as P_i and Q_i. But since $U=0$ when $\nu=\infty$, it follows that $\phi_i(\nu)=0$ when $\nu=\infty$. Hence $\phi_i(\nu)$ is some multiple of $Q_i(\nu) = AQ_i(\nu)$ suppose. It remains to determine A.

Now, $\phi_i(\nu)$ may be developed in a series proceeding by ascending powers of $\frac{1}{\nu}$, as follows.

We have $\quad \frac{1}{\nu-\mu} = \frac{1}{\nu} + \frac{\mu}{\nu^2} + \ldots + \frac{\mu^i}{\nu^{i+1}} + \ldots.$

and also $\quad = \phi_0(\nu)\,P_0(\mu) + \phi_1(\nu)\,P_1(\mu) + \ldots + \phi_i(\nu)\,P_i(\mu) + \ldots$

Now, by Chap. II. Art. 17, we see that, if m be any integer greater than i, the coefficient of P_i in μ^m is

$$(2i+1)\frac{(m-i+2)(m-i+4)\ldots(m-1)}{(m+i+1)(m+i-1)\ldots(m+4)(m+2)} \text{ if } i \text{ be odd,}$$

$$\text{and } (2i+1)\frac{(m-i+2)(m-i+4)\ldots m}{(m+i+1)(m+i-1)\ldots(m+3)(m+1)} \text{ if } i \text{ be even,}$$

$m-i$ being always even.

Hence, writing for m successively i, $i+2$, $i+4$, ... we get

$$\phi_i(\nu) = (2i+1)\left\{\frac{2.4\ldots(i-1)}{(2i+1)(2i-1)\ldots(i+2)}\frac{1}{\nu^{i+1}} + \frac{4.6\ldots(i+1)}{(2i+3)(2i+1)\ldots(i+4)}\frac{1}{\nu^{i+3}} + \frac{6.8\ldots(i+3)}{(2i+5)(2i+3)\ldots(i+6)}\frac{1}{\nu^{i+5}} + \ldots\right\} \text{ if } i \text{ be odd,}$$

and $=(2i+1)\left\{\frac{2.4\ldots i}{(2i+1)(2i-1)\ldots(i+1)}\frac{1}{\nu_i^{+1}}\right.$

$$+\frac{4.6\ldots(1+2)}{(2i+3)(2i+1)\ldots(i+3)}\frac{1}{\nu^{i+3}}$$

$$\left.+\frac{6.8\ldots(i+4)}{(2i+5)(2i+3)\ldots(i+5)}\frac{1}{\nu^{i+5}}+\ldots\right\}\text{ if } i \text{ be even.}$$

Now, recurring to the equation

$$Q(\nu)=P_i(\nu)\int_\nu^\infty\frac{d\mu}{P_i(\mu)^2(\mu^2-1)},$$

we see that, if $Q_i(\nu)$ be developed in a series of ascending powers of $\frac{1}{\nu}$, the first term will be $\frac{1}{C(2i+1)\nu^{i+1}}$, where C is the coefficient of μ^i in the development of $P_i(\mu)$;

that is $C=\frac{(i+2)(i+4)\ldots(2i-1)}{2.4.6\ldots(i-1)}$ if i be odd,

and $=\frac{(i+1)(i+3)(i+5)\ldots(2i-1)}{2.4.6\ldots i}$ if i be even.

Hence the first term in the development of $Q_i(\nu)$ is

$$=\frac{2.4.6\ldots(i-1)}{(i+2)(i+4)\ldots(2i-1)(2i+1)}\text{ if } i \text{ be odd,}$$

and $=\frac{2.4.6\ldots i}{(i+1)(i+3)\ldots(2i-1)(2i+1)}$ if i be even,

which is the same as the first term of the development of $P_i(\nu)$, divided by $\frac{1}{2i+1}$.

Hence $A=2i+1$, and we have

$$\frac{1}{\nu-\mu}=Q_0(\nu)P_0(\mu)+3Q_1(\nu)P_1(\mu)+5Q_2(\nu)P_2(\mu)+\ldots$$

3. The expression for Q_i may be thrown into a more convenient form, by introducing into the numerator and de-

nominator of the coefficient of each term, the factor necessary to make the numerator the product of i consecutive integers. We shall thus make the denominator the product of i consecutive odd integers, and may write

$$Q_i(\nu) = \frac{1.2.3\dots i}{1.3.5\dots(2i+1)}\frac{1}{\nu^{i+1}} + \frac{3.4.5\dots(i+2)}{3.5.7\dots(2i+3)}\frac{1}{\nu^{i+3}}$$

$$+ \frac{5.6.7\dots(i+4)}{5.7.9\dots(2i+5)}\frac{1}{\nu^{i+5}} + \dots$$

$$+ \frac{(2k+1)(2k+2)\dots(i+2k)}{(2k+1)(2k+3)\dots(2i+2k+1)}\frac{1}{\nu^{i+2k-1}} + \dots.$$

whether i be odd or even.

4. We shall not enter into a full discussion of the properties of Zonal Harmonics of the Second Kind. They will be found very completely treated by Heine, in his *Handbuch der Kugelfunctionen.* We will however, as an example, investigate the expression for $\frac{dQ_i}{d\nu}$ in terms of Q_{i+1}, $Q_{i+3}\dots$

Recurring to the equation

$$\frac{1}{\nu-\mu} = Q_0(\nu)P_0(\mu) + 3Q_1(\nu)P_1(\mu) + \dots$$

$$+ (2i+1)5Q_i(\nu)P_i(\mu) + \dots$$

we see that

$$\frac{d}{d\mu}\frac{1}{\nu-\mu} = Q_0(\nu)\frac{dP_0(\mu)}{d\mu} + 3Q_1(\nu)\frac{dP_1(\mu)}{d\mu} + \dots$$

$$+ (2i+1)5Q_i(\nu)\frac{dP_i(\mu)}{d\mu} + (2i+3)Q_{i+1}(\nu)\frac{dP_{i+1}(\mu)}{d\mu} + \dots.$$

Now we have seen (Chap. II. Art. 22) that

$$\frac{dP_i(\mu)}{d\mu} = (2i-i)P_{i-1}(\mu) + (2i-5)P_{i-3}(\mu) + \dots$$

Hence $$\frac{dP_{i+1}(\mu)}{d\mu} = (2i+1)P_i(\mu) + (2i-3)P_{i-2}(\mu) + \dots$$

$$\frac{dP_{i+3}(\mu)}{d\mu} = (2i+5)\, P_{i+2}(\mu) + (2i+1)\, P_i(\mu) + \ldots$$

$$\frac{dP_{i+5}(\mu)}{d\mu} = (2i+9)\, P_{i+4}(\mu) + (2i+5)\, P_{i+2}(\mu)$$
$$+ (2i+1)\, P_i(\mu) + \ldots$$

And therefore the coefficient of $P_i(\mu)$ in the expansion of $\frac{d}{d\mu}\frac{1}{\nu-\mu}$ is

$$(2i+1)\,\{(2i+3)\, Q_{i+1}(\nu) + (2i+7)\, Q_{i+3}(\nu) + (2i+11)\, Q_{i+5}(\nu) + \ldots\}.$$

Again,

$$\frac{d}{d\nu}\frac{1}{\nu-\mu} = \frac{dQ_0(\nu)}{d\nu} P_0(\mu) + 3\frac{dQ_1(\nu)}{d\nu} P_1(\mu) + \ldots$$
$$+ (2i+1)\frac{dQ_i(\nu)}{d\nu} P_i(\mu).$$

And
$$\frac{d}{d\nu}\frac{1}{\nu-\mu} + \frac{d}{d\mu}\frac{1}{\nu-\mu} = 0.$$

Hence, comparing coefficients of $P_i(\mu)$,

$$\frac{d \,.\, Q_i(\nu)}{d\nu} = -(2i+3)\, Q_{i+1}(\nu) - (2i+7)\, Q_{i+3}(\nu)$$
$$-(2i+11)\, Q_{i+5}(\nu) - \ldots$$

Hence it follows that

$$\frac{d \,.\, Q_i(\nu)}{d\nu} - \frac{d \,.\, Q_{i+2}(\nu)}{d\nu} = -(2i+3)\, Q_{i+1}(\nu),$$

and therefore that

$$\int_\nu^\infty Q_{i+1}(\nu)\, d\nu = \frac{1}{2i+3}\{Q_i(\nu) - Q_{i+2}(\nu)\}.$$

5. By similar reasoning to that by which the existence of Tesseral Harmonics was established, we may prove that there is a system of functions, which may be called Tesseral Harmonics of the Second Kind, derived from $T_i^{(\sigma)}$ in the same

manner as Q_i is derived from P_i. The general type of such expressions will be

$$T_i^{(\sigma)}(\nu)\int_\nu^\infty \frac{d\mu}{\overline{T_i^{(\sigma)}(\mu)}|^2\,(\mu^2-1)},$$

and this when multiplied by $\cos\sigma\phi$ or $\sin\sigma\phi$, will give an expression satisfying the differential equation

$$\left\{(1-\mu^2)\frac{d}{d\mu}\right\}^2 U + \{i(i+1)(1-\mu^2)-\sigma^2\}\,U=0,$$

and which may be called the Tesseral Harmonic of the second kind, of the degree i and order σ.

CHAPTER VI.

ELLIPSOIDAL AND SPHEROIDAL HARMONICS.

1. THE characteristic property of Spherical Harmonics is thus stated by Thomson and Tait (p. 400, Art. 537).

"A spherical harmonic distribution of density on a spherical surface produces a similar and similarly placed spherical harmonic distribution of potential over every concentric spherical surface through space, external and internal."

The object of the present chapter is to establish the existence of certain functions which possess an analogous property for an ellipsoid. They have been treated of by Lamé, in his *Leçons sur les fonctions inverses des transcendantes et les fonctions isothermes*, and were virtually introduced by Green, in his memoir *On the Determination of the Exterior and Interior Attractions of Ellipsoids of Variable Densities*, (Transactions of the Cambridge *Philosophical Society*, 1835). We shall consider them both as functions of the elliptic co-ordinates (as Lamé has done) and also as functions of the ordinary rectangular co-ordinates; and after investigating some of their more important general properties, shall proceed to a more detailed discussion of the forms which they assume, when the ellipsoid is a surface of revolution.

2. For this purpose, it will be necessary to transform the equation

$$\frac{d^2V}{dx^2}+\frac{d^2V}{dy^2}+\frac{d^2V}{dz^2}=0, \text{ or } \nabla^2V=0,$$

into its equivalent, when the elliptic co-ordinates ϵ, v, v' are taken as independent variables. If a, b, c be the semiaxes of the ellipsoid, the two sets of independent variables are connected by the relations

$$\frac{x^2}{a^2+\epsilon}+\frac{y^2}{b^2+\epsilon}+\frac{z^2}{c^2+\epsilon}=1,\quad \frac{x^2}{a^2+\upsilon}+\frac{y^2}{b^2+\upsilon}+\frac{z^2}{c^2+\upsilon}=1,$$

$$\frac{x^2}{a^2+\upsilon'}+\frac{y^2}{b^2+\upsilon'}+\frac{z^2}{c^2+\upsilon'}=1.$$

Thus $a^2+\epsilon$, $b^2+\epsilon$, $c^2+\epsilon$ are the squares on the semiaxes of the confocal ellipsoid passing through the point x, y, z.

$a^2+\upsilon$, $b^2+\upsilon$, $c^2+\upsilon$, the squares on the semiaxes of the confocal hyperboloid of one sheet.

$a^2+\upsilon'$, $b^2+\upsilon'$, $c^2+\upsilon'$, the squares on the semiaxes of the confocal hyperboloid of two sheets.

Thus, ϵ is positive if the point x, y, z be external to the given ellipsoid, negative if it be internal.

And, if a^2 be the greatest, c^2 the least, of the quantities a^2, b^2, c^2,

ϵ	will lie	between	$-c^2$	and	∞,
υ	"	"	$-b^2$	"	$-c^2$,
υ'	"	"	$-a^2$	"	$-b^2$.

3. Now $\dfrac{d^2V}{dx^2}+\dfrac{d^2V}{dy^2}+\dfrac{d^2V}{dz^2}=0$ is the condition that

$$\iiint\left\{\left(\frac{dV}{dx}\right)^2+\left(\frac{dV}{dy}\right)^2+\left(\frac{dV}{dz}\right)^2\right\}dx\,dy\,dz,$$

taken throughout a certain region of space, should be a minimum. In the memoir by Green, above referred to, this expression is transformed into its equivalent in terms of a new system of independent variables, and the methods of the Calculus of Variations are then applied to make the resulting expression a minimum. We shall adopt a direct mode of transformation, as follows:

Suppose α, β, γ to be three functions of x, y, z, such that

$$\nabla^2\alpha=0,\quad \nabla^2\beta=0,\quad \nabla^2\gamma=0\ldots\ldots\ldots\ldots\ldots(1),$$

such also that the three families of surfaces represented by the equations $\alpha=$ constant, $\beta=$ constant, $\gamma=$ constant, intersect each other everywhere at right angles, i.e. such that

$$\frac{d\beta}{dx}\frac{d\gamma}{dx}+\frac{d\beta}{dy}\frac{d\gamma}{dy}+\frac{d\beta}{dz}\frac{d\gamma}{dz}=0, \quad \frac{d\gamma}{dx}\frac{d\alpha}{dx}+\frac{d\gamma}{dy}\frac{d\alpha}{dy}+\frac{d\gamma}{dz}\frac{d\alpha}{dz}=0,$$

$$\frac{d\alpha}{dx}\frac{d\beta}{dx}+\frac{d\alpha}{dy}\frac{d\beta}{dy}+\frac{d\alpha}{dz}\frac{d\beta}{dz}=0 \quad \ldots\ldots\ldots\ldots\ldots(2).$$

Then

$$\frac{dV}{dx}=\frac{dV}{d\alpha}\frac{d\alpha}{dx}+\frac{dV}{d\beta}\frac{d\beta}{dx}+\frac{dV}{d\gamma}\frac{d\gamma}{dx},$$

$$\frac{d^2V}{dx^2}=\frac{d^2V}{d\alpha^2}\left(\frac{d\alpha}{dx}\right)^2+\frac{d^2V}{d\beta^2}\left(\frac{d\beta}{dx}\right)^2+\frac{d^2V}{d\gamma^2}\left(\frac{d\gamma}{dx}\right)^2$$

$$+2\frac{d^2V}{d\beta d\gamma}\frac{d\beta}{dx}\frac{d\gamma}{dx}+2\frac{d^2V}{d\gamma d\alpha}\frac{d\gamma}{dx}\frac{d\alpha}{dx}+2\frac{d^2V}{d\alpha d\beta}\frac{d\alpha}{dx}\frac{d\beta}{dx}$$

$$+\frac{dV}{d\alpha}\frac{d^2\alpha}{dx^2}+\frac{dV}{d\beta}\frac{d^2\beta}{dx^2}+\frac{dV}{d\gamma}\frac{d^2\gamma}{dx^2}.$$

$\frac{d^2V}{dy^2}$ and $\frac{d^2V}{dz^2}$ being similarly formed, we see that, when the three expressions are added together, the terms involving $\frac{dV}{d\alpha}, \frac{dV}{d\beta}, \frac{dV}{d\gamma}$ will disappear by the conditions (1), and those involving $\frac{d^2V}{d\beta d\gamma}, \frac{d^2V}{d\gamma d\alpha}, \frac{d^2V}{d\alpha d\beta}$ by the conditions (2). Hence

$$\nabla^2V=\frac{d^2V}{d\alpha^2}\left\{\left(\frac{d\alpha}{dx}\right)^2+\left(\frac{d\alpha}{dy}\right)^2+\left(\frac{d\alpha}{dz}\right)^2\right\}$$

$$+\frac{d^2V}{d\beta^2}\left\{\left(\frac{d\beta}{dx}\right)^2+\left(\frac{d\beta}{dy}\right)^2+\left(\frac{d\beta}{dz}\right)^2\right\}$$

$$+\frac{d^2V}{d\gamma^2}\left\{\left(\frac{d\gamma}{dx}\right)^2+\left(\frac{d\gamma}{dy}\right)^2+\left(\frac{d\gamma}{dz}\right)^2\right\}.$$

4. Now, let

$$\alpha=\int_{\epsilon}^{\infty}\frac{d\psi}{\{(a^2+\psi)(b^2+\psi)(c^2+\psi)\}^{\frac{1}{2}}},$$

$$\beta^{*}=\int_{v}^{-c^2}\frac{d\psi}{\{(a^2+\psi)(b^2+\psi)(c^2+\psi)\}^{\frac{1}{2}}},$$

$$\gamma=\int_{v'}^{-b^2}\frac{d\psi}{\{(a^2+\psi)(b^2+\psi)(c^2+\psi)\}^{\frac{1}{2}}}.$$

All these expressions satisfy the conditions (1), for α is the potential of a homogeneous ellipsoidal shell, of proper density, at an external point, and β and γ possess the same analytical properties.

Again, α is independent of v and v', and is therefore constant when ϵ is constant. Similarly β is constant when v is constant, and γ is constant when v' is constant. Hence α, β, γ satisfy the conditions (2).

Now

$$\left(\frac{d\lambda}{dx}\right)^2+\left(\frac{d\lambda}{dy}\right)^2+\left(\frac{d\lambda}{dz}\right)^2$$

$$=\frac{1}{(a^2+\epsilon)(b^2+\epsilon)(c^2+\epsilon)}\left\{\left(\frac{d\epsilon}{dx}\right)^2+\left(\frac{d\epsilon}{dy}\right)^2+\left(\frac{d\epsilon}{dz}\right)^2\right\}.$$

And $$\frac{x^2}{a^2+\epsilon}+\frac{y^2}{b^2+\epsilon}+\frac{z^2}{c^2+\epsilon}=1.$$

$$\therefore\ \left\{\frac{x^2}{(a^2+\epsilon)^2}+\frac{y^2}{(b^2+\epsilon)^2}+\frac{z^2}{(c^2+\epsilon)^2}\right\}\frac{d\epsilon}{dx}=\frac{2x}{a^2+\epsilon},$$

with similar expressions for $\dfrac{d\epsilon}{dy}$ and $\dfrac{d\epsilon}{dz}$. Hence, squaring and adding,

$$\left\{\frac{x^2}{(a^2+\epsilon)^2}+\frac{y^2}{(b^2+\epsilon)^2}+\frac{z^2}{(c^2+\epsilon)^2}\right\}\left\{\left(\frac{d\epsilon}{dx}\right)^2+\left(\frac{d\epsilon}{dy}\right)^2+\left(\frac{d\epsilon}{dz}\right)^2\right\}=4.$$

But from the equations

$$\frac{x^2}{a^2+\epsilon}+\frac{y^2}{b^2+\epsilon}+\frac{z^2}{c^2+\epsilon}=1,\quad \frac{x^2}{a^2+v}+\frac{y^2}{b^2+v}+\frac{z^2}{c^2+v}=1,$$

* β is a purely imaginary quantity. We may, if we please, write $\sqrt{-1}\beta'$ for β.

$$\frac{x^2}{a^2+v'}+\frac{y^2}{b^2+v'}+\frac{z^2}{c^2+v'}=1,$$

we deduce

$$1-\frac{x^2}{a^2+\omega}-\frac{y^2}{b^2+\omega}-\frac{z^2}{c^2+\omega}=\frac{(\omega-\epsilon)(\omega-v)(\omega-v')}{(\omega+a^2)(\omega+b^2)(\omega+c^2)},$$

ω being any quantity whatever. For this expression is of 0 dimensions in ω, ϵ, v, v', it vanishes when $\omega=\epsilon$, v, or v', and for those values of ω only, it becomes infinite when $\omega=-a^2$, $-b^2$, or $-c^2$, and for those values of ω only, and it is $=1$ when $\omega=\infty$.

From this, multiplying by $a^2+\omega$, and then putting $\omega=-a^2$, we deduce

$$x^2=\frac{(\epsilon+a^2)(v+a^2)(v'+a^2)}{(a^2-b^2)(a^2-c^2)},$$

a result which will be useful hereafter.

Again, differentiating with respect to ω, and then putting $\omega=\epsilon$,

$$\frac{x^2}{(a^2+\epsilon)^2}+\frac{y^2}{(b^2+\epsilon)^2}+\frac{z^2}{(c^2+\epsilon)^2}=\frac{(\epsilon-v)(\epsilon-v')}{(\epsilon+a^2)(\epsilon+b^2)(\epsilon+c^2)},$$

$$\therefore \left(\frac{d\epsilon}{dx}\right)^2+\left(\frac{d\epsilon}{dy}\right)^2+\left(\frac{d\epsilon}{dz}\right)^2=4\frac{(\epsilon+a^2)(\epsilon+b^2)(\epsilon+c^2)}{(\epsilon-v)(\epsilon-v')},$$

$$\therefore \left(\frac{d\alpha}{dx}\right)^2+\left(\frac{d\alpha}{dy}\right)^2+\left(\frac{d\alpha}{dz}\right)^2=\frac{4}{(\epsilon-v)(\epsilon-v')},$$

$$\therefore \nabla^2 V=\frac{4}{(v-v')(v'-\epsilon)(\epsilon-v)}\left\{(v'-v)\frac{d^2V}{d\alpha^2}+(\epsilon-v')\frac{d^2V}{d\beta^2}+(v-\epsilon)\frac{d^2V}{d\gamma^2}\right\}.$$

The equation $\nabla^2 V=0$ is thus transformed into

$$(v'-v)\frac{d^2V}{d\alpha^2}+(\epsilon-v')\frac{d^2V}{d\beta^2}+(v-\epsilon)\frac{d^2V}{d\gamma^2}=0,$$

or
$$(v'-v)\left[\{(\epsilon+a^2)(\epsilon+b^2)(\epsilon+c^2)\}^{\frac{1}{2}}\frac{d}{d\epsilon}\right]^2 V$$
$$+(\epsilon-v')\left[\{(v+a^2)(v+b^2)(v+c^2)\}^{\frac{1}{2}}\frac{d}{dv}\right]^2 V$$
$$+(v-\epsilon)\left[\{(v'+a^2)(v'+b^2)(v'+c^2)\}^{\frac{1}{2}}\frac{d}{dv'}\right]^2 V.$$

5. A class of integrals of this equation, presenting a close analogy to spherical harmonic functions, may be investigated in the following manner. Suppose E to be a function of ϵ, satisfying the equation

$$\left[\{(\epsilon+a^2)(\epsilon+b^2)(\epsilon+c^2)\}^{\frac{1}{2}}\frac{d}{d\epsilon}\right]^2 V=(m\epsilon+r)E,$$

m and r being any constants.

Then, if H and H' be the forms which this function assumes when v and v' are respectively substituted for ϵ, the equation $\nabla^2 V=0$ will be satisfied by $V=EHH'$.

6. We will first investigate the form of the function denoted by E, on the supposition that E is a rational integral function of ϵ of the degree n, represented by

$$\epsilon^n+np_1\epsilon^{n-1}+\frac{n(n-1)}{1.2}p_2\epsilon^{n-2}+\ldots+p_n.$$

We see that

$$\left[\{(\epsilon+a^2)(\epsilon+b^2)(\epsilon+c^2)\}^{\frac{1}{2}}\frac{d}{d\epsilon}\right]^2\left\{\epsilon^n+np_1\epsilon^{n-1}+\frac{n(n-1)}{1.2}p_2\epsilon^{n-2}+\ldots\ldots+p_n\right\}$$

$$=n\left[(n-1)(\epsilon+a^2)(\epsilon+b^2)(\epsilon+c^2)\left\{\epsilon^{n-2}+(n-2)p_1\epsilon^{n-3}+\frac{(n-2)(n-3)}{1.2}p_2\epsilon^{n-4}+\ldots+p_{n-2}\right\}\right.$$
$$\left.+\frac{(\epsilon+b^2)(\epsilon+c^2)+(\epsilon+c^2)(\epsilon+a^2)+(\epsilon+a^2)(\epsilon+b^2)}{2}\left\{\epsilon^{n-1}+(n-1)p_1\epsilon^{n-2}+\frac{(n-1)(n-2)}{1.2}p_2\epsilon^{n-3}+\ldots+p_{n-1}\right\}\right].$$

Hence writing

$$(\epsilon+a^2)(\epsilon+b^2)(\epsilon+c^2)=\epsilon^3+3f_1\epsilon^2+3f_2\epsilon+f_3,$$

we see that

$$n\left[(n-1)(\epsilon^3+3f_1\epsilon^2+3f_2\epsilon+f_3)\left\{\epsilon^{n-2}+(n-2)p_1\epsilon^{n-3}\right.\right.$$
$$\left.+\frac{(n-2)(n-3)}{1.2}p_2\epsilon^{n-4}+\ldots+p_{n-2}\right\}$$
$$+\frac{3}{2}(\epsilon^2+2f_1\epsilon+f_2)\left\{\epsilon^{n-1}+(n-1)p_1\epsilon^{n-2}+\frac{(n-1)(n-2)}{1.2}p_2\epsilon^{n-3}\right.$$
$$\left.\left.+\ldots+p_{n-1}\right\}\right]$$
$$=(m\epsilon+r)\left\{\epsilon^n+np_1\epsilon^{n-1}+\frac{n(n-1)}{1.2}p_2\epsilon^{n-2}+\ldots+p_n\right\}.$$

Hence, equating coefficients of like powers of ϵ, we get

$$n\left(n+\frac{1}{2}\right)=m,$$

$$n\left[(n-1)\{(n-2)p_1+3f_1\}+\frac{3}{2}\{(n-1)p_1+2f_1\}\right]=nmp_1+r,$$

$$n\left[(n-1)\left\{\frac{(n-2)(n-3)}{1.2}p_2+3(n-2)f_1p_1+3f_2\right\}\right.$$
$$\left.+\frac{3}{2}\left\{\frac{(n-1)(n-2)}{1.2}p_2+2(n-1)f_1p_1+f_2\right\}\right]$$
$$=\frac{n(n-1)}{1.2}mp_2+nrp_1$$

$$\ldots\ldots\ldots=\ldots\ldots\ldots$$

$$n\left\{(n-1)f_3p_{n-2}+\frac{3}{2}f_2p_{n-1}\right\}=rp_n,$$

or, as they may be more simply written,

$$n\left(n+\frac{1}{2}\right)=m,$$

$$n\left\{(n-1)\left(n-\frac{1}{2}\right)p_1^2+3nf_1\right\}=nmp_1+r,$$

$$n\left\{\frac{(n-1)(n-2)}{1\,.\,2}\left(n-\frac{3}{2}\right)p_2+3(n-1)^2f_1p_1+3\left(n-\frac{1}{2}\right)f_2\right\}$$

$$=\frac{n(n-1)}{1\,.\,2}mp_2+nrp_1$$

$$\ldots\ldots\ldots=\ldots\ldots\ldots$$

$$n\left\{(n-1)f_3p_{n-2}+\frac{3}{2}f_2p_{n-1}\right\}=rp_n.$$

It thus appears that p_1 is a rational function of r of the first degree, p_2 of the second, p_n of the nth, and when the letters $p_1, p_2 \ldots p_n$ have been eliminated, the resulting equation for the determination of r will be of the $(n+1)^{\text{th}}$ degree. Each of the letters $p_1, p_2 \ldots p_n$ will have one determinate value corresponding to each of these values of r; and we have seen that $m=n\left(n+\frac{1}{2}\right)$. There will therefore be $(n+1)$ values of E, each of which is a rational integral expression of the nth degree, n being any positive integer.

7. But there will also be values of E, of the nth degree, of the form

$$(\epsilon+b^2)^{\frac{1}{2}}(\epsilon+c^2)^{\frac{1}{2}}\left\{\epsilon^{n-1}+(n-1)q_1\epsilon^{n-2}+\frac{(n-1)(n-2)}{1\,.\,2}q_2\epsilon^{n-3}+\ldots+q_{n-1}\right\}:$$

We thus obtain

$$\{(\epsilon+a^2)(\epsilon+b^2)(\epsilon+c^2)\}^{\frac{1}{2}}\frac{dE}{d\epsilon}$$

$$=(\epsilon+a^2)^{\frac{1}{2}}(\epsilon+b^2)(\epsilon+c^2)(n-1)\left\{\epsilon^{n-2}+(n-2)q_1\epsilon^{n-3}\right.$$

$$\left.+\frac{(n-2)(n-3)}{1\,.\,2}q_2\epsilon^{n-4}+\ldots+q_{n-2}\right\};$$

$$\therefore \left[\{(\epsilon+a^2)(\epsilon+b^2)(\epsilon+c^2)\}^{\frac{1}{2}}\frac{d}{d\epsilon}\right]^2 E$$

$$= \{(\epsilon+a^2)(\epsilon+b^2)(\epsilon+c^2)\}^{\frac{1}{2}}\left[\left\{\frac{1}{2}\frac{(\epsilon+b^2)(\epsilon+c^2)}{(\epsilon+a^2)^{\frac{1}{2}}} + (\epsilon+a^2)^{\frac{1}{2}}(\epsilon+c^2)\right.\right.$$

$$\left. + (\epsilon+a^2)^{\frac{1}{2}}(\epsilon+b^2)\right\}(n-1)\left\{\epsilon^{n-2} + (n-2)\,q_1\epsilon^{n-3}\right.$$

$$\left. + \frac{(n-2)(n-3)}{1\,.\,2}\,q_2\epsilon^{n-4} + \ldots + q_{n-2}\right\}$$

$$+ (\epsilon+a^2)^{\frac{1}{2}}(\epsilon+b^2)(\epsilon+c^2)(n-1)(n-2)\left\{\epsilon^{n-3} + (n-3)\,q_1\epsilon^{n-4}\right.$$

$$\left.\left. + \frac{(n-3)(n-4)}{1\,.\,2}\,q_2\epsilon^{n-5} + \ldots + q_{n-3}\right\}\right].$$

Hence

$$\left\{\frac{1}{2}(\epsilon+b^2)(\epsilon+c^2) + (\epsilon+a^2)(\epsilon+c^2) + (\epsilon+a^2)(\epsilon+b^2)\right\}$$

$$(n-1)\left\{\epsilon^{n-2} + (n-2)\,q_1\epsilon^{n-3} + \frac{(n-2)(n-3)}{1\,.\,2}\,q_2\epsilon^{n-4} + \ldots + q_{n-2}\right\}$$

$$+ (\epsilon+a^2)(\epsilon+b^2)(\epsilon+c^2)(n-1)(n-2)\left\{\epsilon^{n-3} + (n-3)\,q_1\epsilon^{n-4}\right.$$

$$\left. + \frac{(n-3)(n-4)}{1\,.\,2}\,q_2\epsilon^{n-5} + \ldots + q_{n-3}\right\}$$

$$= (m\epsilon + r)\left\{\epsilon^{n-1} + (n-1)\,q_1\epsilon^{n-2}\right.$$

$$\left. + \frac{(n-1)(n-2)}{1\,.\,2}\,q_2\epsilon^{n-3} + \ldots + q_{n-1}\right\};$$

$$\therefore (n-1)\left(\frac{5}{2} + n - 2\right) = m,$$

$$(n-1)\left\{2a^2 + \frac{3}{2}(b^2+c^2) + \frac{5}{2}(n-2)\,q_1\right\}$$

$$+ (n-1)(n-2)\{a^2+b^2+c^2+(n-3)\,q_1\} = (n-1)\,mq_1 + r,$$

$$(n-1)\left\{\left(\frac{b^2c^2}{2} + a^2c^2 + a^2b^2\right)q_{n-2} + (n-2)\,a^2b^2c^2\,q_{n-3}\right\} = rq_{n-1}.$$

By a similar process to that applied above, we shall find that r is determined by an equation of the nth degree, and that $m = (n-1)\left(n - \frac{1}{2}\right)$, and that each of the letters q_1, $q_2 \ldots q_{n-1}$ is a rational function of r. Thus, there will be n solutions of the form

$$(\epsilon + b^2)^{\frac{1}{2}} (\epsilon + c^2)^{\frac{1}{2}} \{\epsilon^{n-1} + (n-1) q_1 \epsilon^{n-2} + \ldots + q_{n-1}\}.$$

There will also be n solutions of a similar form, in which the factors $(\epsilon + c^2)^{\frac{1}{2}}$ $(\epsilon + a^2)^{\frac{1}{2}}$, $(\epsilon + a^2)^{\frac{1}{2}} (\epsilon + b^2)^{\frac{1}{2}}$ are respectively involved. Hence, the total number of solutions of the nth degree will be $4n + 1$.

8. We may now investigate the number of solutions of the degree $n + \frac{1}{2}$, n being any positive integer. These will be of the following forms: three obtained by multiplying a rational integral function of ϵ of the degree n by $(\epsilon + a^2)^{\frac{1}{2}}$, $(\epsilon + b^2)^{\frac{1}{2}}$, $(\epsilon + c^2)^{\frac{1}{2}}$, respectively, and one by multiplying a rational integral function of ϵ of the degree $n - 1$ by the product

$$\{(\epsilon + a^2)(\epsilon + b^2)(\epsilon + c^2)\}^{\frac{1}{2}}.$$

An exactly similar process to that applied above will shew us that there will be $n + 1$ solutions of each of the first three kinds, and n of the fourth. Hence the total number of such solutions will be $3(n+1) + n$, or $4n + 3$, that is $4\left(n + \frac{1}{2}\right) + 1$.

To sum up these results, we may say that the total number of solutions of the nth degree is $4n + 1$, n denoting either a positive integer, or a fraction with an odd numerator, and denominator 2.

Similar forms being obtained for H, H', we may proceed to transform the expression EHH' into a function of x, y, z.

9. Consider first the case in which

$$E = \epsilon^n + np_1\epsilon^{n-1} + \frac{n(n-1)}{1 \cdot 2} p_2\epsilon^{n-2} + \ldots + p_n.$$

Write this under the form

$$E=(\epsilon-\omega_1)(\epsilon-\omega_2)\dots(\epsilon-\omega_n).$$

Then
$$H=(\upsilon-\omega_1)(\upsilon-\omega_2)\dots(\upsilon-\omega_n),$$
$$H'=(\upsilon'-\omega_1)(\upsilon'-\omega_2)\dots(\upsilon'-\omega_n).$$

Hence

$$EHH'=(\epsilon-\omega_1)(\upsilon-\omega_1)(\upsilon'-\omega_1)\dots(\epsilon-\omega_n)(\upsilon-\omega_n)(\upsilon'-\omega_n).$$

Now we have shewn (see Art. 4 of the present Chapter) that $(\epsilon-\omega_1)(\upsilon-\omega_1)(\upsilon'-\omega_1)$

$$=(\omega_1+a^2)(\omega_1+b^2)(\omega_1+c^2)\left(\frac{x^2}{a^2+\omega_1}+\frac{y^2}{b^2+\omega_1}+\frac{z^2}{c^2+\omega_1}-1\right).$$

Each of the factors of EHH' being similarly transformed, we see that EHH' is equal to the continued product of all expressions of the form

$$(\omega+a^2)(\omega+b^2)(\omega+c^2)\left(\frac{x^2}{a^2+\omega}+\frac{y^2}{b^2+\omega}+\frac{z^2}{c^2+\omega}-1\right),$$

the several values of ω being the roots of the equation

$$\omega^n+np_1\omega^{n-1}+\frac{n(n-1)}{1.2}p_2\omega^{n-2}+\dots+p_n=0.$$

As this equation has been already shewn to have $(n+1)$ distinct forms, we obtain $(n+1)$ distinct solutions of the equation $\nabla^2V=0$, each solution being the product of n expressions of the form

$$\frac{x^2}{a^2+\omega}+\frac{y^2}{b^2+\omega}+\frac{z^2}{c^2+\omega}-1.$$

That is, there will be $n+1$ independent solutions of the degree $2n$ in x, y, z, each involving only *even* powers of the variables.

10. To complete the investigation of the number of solutions of the degree $2n$, let us next consider the case in which E

$$=(\epsilon+b^2)^{\frac{1}{2}}(\epsilon+c^2)^{\frac{1}{2}}\left\{\epsilon^{n-1}+(n-1)p_1\epsilon^{n-2}+\frac{(n-1)(n-2)}{1.2}p_2\epsilon^{n-3}+\dots+p_{n-1}\right\}.$$

The object here will be to transform the product

$$(\epsilon + b^2)^{\frac{1}{2}} (v + b^2)^{\frac{1}{2}} (v' + b^2)^{\frac{1}{2}} (\epsilon + c^2)^{\frac{1}{2}} (v + c^2)^{\frac{1}{2}} (v' + c^2)^{\frac{1}{2}},$$

since the other factors will, as already shewn, give rise to the product of $n-1$ expressions of the form

$$\frac{x^2}{a^2+\omega} + \frac{y^2}{b^2+\omega} + \frac{z^2}{c^2+\omega} - 1.$$

Now, by comparison of the value of x^2 given in Art. 4, we see that

$$(\epsilon + b^2)(v + b^2)(v' + b^2)(\epsilon + c^2)(v + c^2)(v' + c^2)$$
$$= (b^2 - c^2)(b^2 - a^2)(c^2 - a^2)(c^2 - b^2)\, y^2 z^2.$$

Hence, we obtain a system of solutions of the form of the product of $(n-1)$ expressions of the form

$$\frac{x^2}{a^2+\omega} + \frac{y^2}{b^2+\omega} + \frac{z^2}{c^2+\omega} - 1,$$

multiplied by yz. Of these there will be n, and an equal number of solutions in which zx, xy, respectively, take the place of yz.

Thus, there will be $4n+1$ solutions of the degree $2n$ in the variables of which $n+1$ are each the product of n expressions of the form

$$\frac{x^2}{a^2+\omega} + \frac{y^2}{b^2+\omega} + \frac{z^2}{c^2+\omega} - 1,$$

n are each the product of $(n-1)$ such expressions, multiplied by yz,

n ... ... zx,

n ... ... xy.

11. We may next proceed to consider the solutions of the degree $2n+1$ in the variables x, y, z.

Consider first the case in which

$$E = (\epsilon + a^2)^{\frac{1}{2}} \left\{\epsilon^n + np_1\epsilon^{n-1} + \frac{n\,.\,(n-1)}{1\,.\,2} p_2\epsilon^{n-2} + \ldots + p_n\right\}.$$

Here the product $(\epsilon + a^2)^{\frac{1}{2}} (\upsilon + a^2)^{\frac{1}{2}} (\upsilon' + a^2)^{\frac{1}{2}}$ will, as just shewn, give rise to a factor x in the product EHH'.

Hence we obtain a system of solutions each of which is the product of n expressions of the form

$$\frac{x^2}{a^2+\omega} + \frac{y^2}{b^2+\omega} + \frac{z^2}{c^2+\omega} - 1,$$

multiplied by x. Of these there will be $n+1$, and an equal number of solutions in which y, z, respectively take the place of the factor x.

Lastly, in the case in which

$$E = (\epsilon + a^2)^{\frac{1}{2}} (\epsilon + b^2)^{\frac{1}{2}} (\epsilon + c^2)^{\frac{1}{2}} \left\{\epsilon^{n-1} + (n-1) p_1 \epsilon^{n-2} + \frac{(n-1)(n-2)}{1.2} p_2 \epsilon^{n-2} + \ldots + p_{n-1}\right\},$$

we see that in EHH' the product

$$(\epsilon+a^2)^{\frac{1}{2}} (\upsilon+a^2)^{\frac{1}{2}} (\upsilon'+a^2)^{\frac{1}{2}} (\epsilon+b^2)^{\frac{1}{2}} (\upsilon+b^2)^{\frac{1}{2}} (\upsilon'+b^2)^{\frac{1}{2}} (\epsilon+c^2)^{\frac{1}{2}} (\upsilon+c^2)^{\frac{1}{2}} (\upsilon'+c^2)^{\frac{1}{2}}$$

will give rise to a factor xyz.

Hence we obtain a system of solutions each of which is the product of $(n-1)$ expressions of the form

$$\frac{x^2}{a^2+\omega} + \frac{y^2}{b^2+\omega} + \frac{z^2}{c^2+\omega} - 1,$$

multiplied by xyz. Of these there will be n.

Thus there will be $4n+3$ solutions of the degree $2n+1$ in the variables, of which

$(n+1)$ are each the product of n expressions of the form

$$\frac{x^2}{a^2+\omega} + \frac{y^2}{b^2+\omega} + \frac{z^2}{c^2+\omega} - 1 \text{ multiplied by } x,$$

$(n+1)$ are each the product of n such expressions, multiplied by y,

$(n+1)$ ... ... z,

n are each the product of $(n-1)$ such expressions, multiplied by xyz.

12. Now an expression of the form $C . EHH'$, C being any arbitrary constant, is an admissible value of the potential at any point *within* the shell $\frac{x^2}{a^2}+\frac{y^2}{b^2}+\frac{z^2}{c^2}=1$. But it is not admissible for the space *without* the shell, since it becomes infinite at an infinite distance. The factor which becomes infinite is clearly E, and we have therefore to enquire whether any form, free from this objection, can be found for this factor. We shall find that forms exist, bearing the same relation to E that zonal harmonics of the second kind bear to those of the first.

Now considering the equation

$$\left[\{(\epsilon+a^2)(\epsilon+b^2)(\epsilon+c^2)\}^{\frac{1}{2}}\frac{d}{d\epsilon}\right]^2 U=(m\epsilon+r)\,U,$$

which we suppose to be satisfied by putting $U=E$, we see that, since it is of the second order, it must admit of another particular integral. To find this, substitute for U, $E\int v d\epsilon$, we then have

$$\left[\{(\epsilon+a^2)(\epsilon+b^2)(\epsilon+c^2)\}^{\frac{1}{2}}\frac{d}{d\epsilon}\right] U$$

$$=\left[\{(\epsilon+a^2)(\epsilon+b^2)(\epsilon+c^2)\}^{\frac{1}{2}}\frac{d}{d\epsilon}\right] E.\int v d\epsilon$$

$$+\{(\epsilon+a^2)(\epsilon+b^2)(\epsilon+c^2)\}^{\frac{1}{2}}Ev;$$

$$\therefore \left[\{(\epsilon+a^2)(\epsilon+b^2)(\epsilon+c^2)\}^{\frac{1}{2}}\frac{d}{d\epsilon}\right]^2 U$$

$$=\left[\{(\epsilon+a^2)(\epsilon+b^2)(\epsilon+c^2)\}^{\frac{1}{2}}\frac{d}{d\epsilon}\right]^2 E.\int v d\epsilon$$

$$+(\epsilon+a^2)(\epsilon+b^2)(\epsilon+c^2)\frac{dE}{d\epsilon}.v$$

$$+\frac{1}{2}\{(\epsilon+b^2)(\epsilon+c^2)+(\epsilon+c^2)(\epsilon+a^2)+(\epsilon+a^2)(\epsilon+b^2)\}\,Ev$$

$$+(\epsilon+a^2)(\epsilon+b^2)(\epsilon+c^2)\left(\frac{dE}{d\epsilon}.v+E\frac{dv}{d\epsilon}\right).$$

Now, since by supposition, the equation for the determination of U is satisfied by putting $U=E$, it follows that when $E\int v d\epsilon$ is substituted for U, the terms involving $\int v d\epsilon$ will cancel each other, and the equation for the determination of v will be reduced to

$$E\frac{dv}{d\epsilon}+\left\{2\frac{dE}{d\epsilon}+\frac{1}{2}\left(\frac{1}{\epsilon+a^2}+\frac{1}{\epsilon+b^2}+\frac{1}{\epsilon+c^2}\right)E\right\}v=0,$$

or
$$\frac{1}{v}\frac{dv}{d\epsilon}+\frac{2}{E}\frac{dE}{d\epsilon}+\frac{1}{2}\left(\frac{1}{\epsilon+a^2}+\frac{1}{\epsilon+b^2}+\frac{1}{\epsilon+c^2}\right)=0;$$

whence
$$\log v+2\log E+\log\{(\epsilon+a^2)(\epsilon+b^2)(\epsilon+c^2)\}^{\frac{1}{2}}$$
$$=\log v_0+2\log E_0+\log abc,$$

v_0 and E_0 being the values of v and E, corresponding to $\epsilon=0$.

Hence
$$v=v_0\frac{E_0^2}{E^2}\frac{abc}{\{(\epsilon+a^2)(\epsilon+b^2)(\epsilon+c^2)\}^{\frac{1}{2}}};$$

$$\therefore E\int v d\epsilon=v_0E_0^2\,abc\,.\,E\int\frac{d\epsilon}{E^2\{(\epsilon+a^2)(\epsilon+b^2)(\epsilon+c^2)\}^{\frac{1}{2}}}.$$

We may therefore take, as a value of the potential at any external point,

$$V=v_0E_0^2\,abc\,EHH'\int_\epsilon^\infty\frac{d\epsilon}{E^2\{(\epsilon+a^2)(\epsilon+b^2)(\epsilon+c^2)\}^{\frac{1}{2}}}.$$

For this obviously vanishes when $\epsilon=\infty$. It remains so to determine v_0 that this value shall, at the surface of the ellipsoid, be equal to the value $C.EHH'$, already assumed for an internal point. This gives

$$C=v_0\,.\,E_0^2\,abc\int_0^\infty\frac{d\epsilon}{E^2\{(\epsilon+a^2)(\epsilon+b^2)(\epsilon+c^2)\}^{\frac{1}{2}}}.$$

Hence, putting $v_0\,.\,E_0^2\,.\,abc=V_0$, we see that to the value of the potential

$$V_0\,EHH'\int_0^\infty\frac{d\epsilon}{E^2\{(\epsilon+a^2)(\epsilon+b^2)(\epsilon+c^2)\}^{\frac{1}{2}}},$$

for any internal point, corresponds the value

$$V_0 EHH' \int_\epsilon^\infty \frac{d\epsilon}{E^2 \{(\epsilon + a^2)(\epsilon + b^2)(\epsilon + c^2)\}^{\frac{1}{2}}},$$

for any external point.

13. We proceed to investigate the law of distribution of density of attracting matter over the surface of the ellipsoid, corresponding to such a distribution of potential.

Now, generally, if δn be the thickness of a shell, ρ its volume density, the difference between the normal components of the attraction of the shell on two particles, situated close to the shell, on the same normal, one within and the other without will be $4\pi\rho\delta n$. This is the attraction of the shell on the *outer* particle, minus the attraction on the *inner* particle.

But the normal component of the attraction on the outer particle estimated inwards is $-\dfrac{dV}{dn}$.

And, if V' denote the potential of the shell on an internal particle, the normal component of the attraction on it estimated inwards is $-\dfrac{dV'}{dn}$.

Hence
$$4\pi\rho\delta n = \frac{dV'}{dn} - \frac{dV}{dn}.$$

Now
$$\frac{dV}{dn} = \frac{dV}{dx}\cdot\frac{dx}{dn} + \frac{dV}{dy}\frac{dy}{dn} + \frac{dV}{dz}\frac{dz}{dn}.$$

And $\dfrac{dx}{dn}$ is the cosine of the inclination of the normal at the point x, y, z to the axis of x, and is therefore generally equal to $e\dfrac{x}{a^2+\epsilon}$, e denoting the perpendicular from the centre on the tangent plane to the surface

$$\frac{x^2}{a^2+\epsilon} + \frac{y^2}{b^2+\epsilon} + \frac{z^2}{c^2+\epsilon} = 1.$$

And we have shewn that

$$x^2 = \frac{(a^2+\epsilon)(a^2+v)(a^2+v')}{(a^2-b^2)(a^2-c^2)},$$

whence

$$\frac{2}{x}\frac{dx}{d\epsilon} = \frac{1}{a^2+\epsilon},$$

or

$$\frac{x}{a^2+\epsilon} = 2\frac{dx}{d\epsilon};$$

$$\therefore \frac{dx}{dn} = 2e\frac{dx}{d\epsilon}.$$

Similarly $\quad \dfrac{dy}{dn} = 2e\dfrac{dy}{d\epsilon}, \quad \dfrac{dz}{dn} = 2e\dfrac{dz}{d\epsilon},$

$$\therefore \frac{dV}{dn} = 2e\left(\frac{dV}{dx}\frac{dx}{d\epsilon} + \frac{dV}{dy}\frac{dy}{d\epsilon} + \frac{dV}{dz}\frac{dz}{d\epsilon}\right) = 2e\frac{dV}{d\epsilon}.$$

Similarly $\quad \dfrac{dV'}{dn} = 2e\dfrac{dV'}{d\epsilon}.$

Now $\quad V' = V_0 . EHH'\displaystyle\int_0^\infty \frac{d\epsilon}{E^2\{(\epsilon+a^2)(\epsilon+b^2)(\epsilon+c^2)\}^{\frac{1}{2}}};$

$$\therefore \frac{dV'}{d\epsilon} = V_0 . HH'\frac{dE}{d\epsilon}\int_0^\infty \frac{d\epsilon}{E^2\{(\epsilon+a^2)(\epsilon+b^2)(\epsilon+c^2)\}^{\frac{1}{2}}}.$$

And $\quad V = V_0 . EHH'\displaystyle\int_\epsilon^\infty \frac{d\epsilon}{E^2\{(a^2+\epsilon)(b^2+\epsilon)(c^2+\epsilon)\}^{\frac{1}{2}}};$

therefore, generally,

$$\frac{dV}{d\epsilon} = V_0 . HH'\frac{dE}{d\epsilon}\int_\epsilon^\infty \frac{d\epsilon}{E^2\{(a^2+\epsilon)(b^2+\epsilon)(c^2+\epsilon)\}^{\frac{1}{2}}}$$

$$- V_0 . EHH'\frac{1}{E^2\{(a^2+\epsilon)(b^2+\epsilon)(c^2+\epsilon)\}^{\frac{1}{2}}}.$$

But, when the attracted particle is in the immediate neighbourhood of the surface, $\epsilon = 0$. Hence, the first line of the value of $\frac{dV}{d\epsilon}$ becomes identical with the value of $\frac{dV'}{d\epsilon}$, and we have

$$\frac{dV'}{d\epsilon} - \frac{dV}{d\epsilon} = V_0 \frac{HH'}{E_0} \frac{1}{abc},$$

E_0 denoting the value which E assumes, when $\epsilon = 0$.

Hence, $$4\pi\rho\delta n = 2eV_0 \frac{HH'}{E_0} \frac{1}{abc}.$$

But δn, being the thickness of the shell, is proportional to e, and we may therefore write $\frac{e}{\delta n} = \frac{a}{\delta a}$, δa being the thickness of the shell at the extremity of the greatest axis;

$$\therefore \rho = \frac{V_0}{2\pi} \frac{a}{\delta a} \frac{1}{abc} \frac{HH'}{E_0},$$

and this is proportional to the value of V corresponding to any specified value of ϵ, since HH' is the only variable factor in either.

Hence functions of the kind which we are now considering possess a property analogous to that of Spherical Harmonics quoted at the beginning of this Chapter. On account of this property, we propose to call them Ellipsoidal Harmonics, and shall distinguish them, when necessary, into surface and solid harmonics, in the same manner as spherical harmonics are distinguished. They are commonly known as Lamé's Functions, having been fully discussed by him in his *Leçons*. The equivalent expressions in terms of x, y, z have been considered by Green in his Memoir mentioned at the beginning of this chapter, and for this reason Professor Cayley in his "Memoir on Prepotentials," read before the Royal Society on June 10, 1875, calls them "Greenians."

We may observe that the factor

$$\frac{1}{4\pi} \frac{a}{\delta a} \frac{1}{abc}$$

is equal to $\frac{1}{4\pi bc\delta a}$, and therefore also to $\frac{1}{4\pi ca\delta b}$ or $\frac{1}{4\pi ab\delta c}$.

Hence, it is equal to

$$\frac{1}{\frac{4\pi}{3}(bc\delta a + ca\delta b + ab\delta c)},$$

or to

$$\frac{1}{\text{volume of shell}};$$

and the potential at any internal point

$$= \tfrac{1}{2} \text{ volume of shell} \times EE_0 . \rho \int_0^\infty \frac{d\epsilon}{E^2 \{(a^2+\epsilon)(b^2+\epsilon)(c^2+\epsilon)\}^{\frac{1}{2}}},$$

and the potential at any external point

$$= \tfrac{1}{2} \text{ volume of shell} \times EE_0 . \rho \int_\epsilon^\infty \frac{d\epsilon}{E^2 \{(a^2+\epsilon)(b^2+\epsilon)(c^2+\epsilon)\}^{\frac{1}{2}}};$$

where for ρ must be substituted its value in terms of v and v'.

14. We will next prove that if V_1, V_2 be two different ellipsoidal harmonics, dS an element of the surface of the ellipsoid, $\iint e\, V_1 V_2\, dS = 0$, the integration being extended all over the surface.

We have generally

$$\iiint (V_1 \nabla^2 V_2 - V_2 \nabla^2 V_1)\, dx\, dy\, dz = \iint \left(V_1 \frac{dV_2}{dn} - V_2 \frac{dV_1}{dn} \right) dS$$

$$= 2 \iint e \left(V_1 \frac{dV_2}{d\epsilon} - V_2 \frac{dV_1}{d\epsilon} \right) dS.$$

And throughout the space comprised within the limits of integration, $\nabla^2 V_1 = 0$, $\nabla^2 V_2 = 0$. Hence

$$\iint e \left(V_1 \frac{dV_2}{d\epsilon} - V_2 \frac{dV_1}{d\epsilon} \right) dS = 0.$$

Now it has been shewn already that V_1, V_2 are each of the form EHH', where E is a function of ϵ only, H the same function of v, H' of v'. We may therefore write

$$V_1 = f_1(\epsilon) f_1(v) f_1(v'),$$

and similarly $$V_2 = f_2(\epsilon) f_2(v) f_2(v').$$

Hence
$$V_1 \frac{dV_2}{d\epsilon} = V_1 V_2 \frac{f_2'(\epsilon)}{f_2(\epsilon)},$$
$$V_2 \frac{dV_1}{d\epsilon} = V_1 V_2 \frac{f_1'(\epsilon)}{f_1(\epsilon)};$$
$$\therefore V_1 \frac{dV_2}{d\epsilon} - V_2 \frac{dV_1}{d\epsilon} = V_1 V_2 \left\{\frac{f_2'(\epsilon)}{f_2(\epsilon)} - \frac{f_1'(\epsilon)}{f_1(\epsilon)}\right\}.$$

Now, all over the surface, $\epsilon = 0$. Hence
$$\iint e\, V_1 V_2 \left\{\frac{f_2'(0)}{f_2(0)} - \frac{f_1'(0)}{f_1(0)}\right\} dS = 0.$$

Hence, unless $\dfrac{f_2'(0)}{f_2(0)} - \dfrac{f_1'(0)}{f_1(0)} = 0$, which cannot happen unless the functions denoted by f_1 and f_2 are identical*, or only differ by a numerical factor, we must have
$$\iint e\, V_1 V_2\, dS = 0.$$

Now e is proportional to the thickness of the shell at any point. Calling this thickness δe, we have therefore
$$\iint \delta e\, V_1 V_2\, dS = 0.$$

Hence, adding together the results obtained by integrating successively over a continuous series of such surfaces, we get
$$\iiint V_1 V_2\, dx dy dz = 0;$$
V_1, V_2 now representing solid ellipsoidal harmonics, and the integration extending throughout the whole space comprised within the ellipsoid.

* This may be shewn more rigorously by integrating through the space bounded by two confocal ellipsoids, defined by the values λ and μ of ϵ. We then get, as in the text,
$$\iint e\, V_1 V_2 \left\{\frac{f_2'(\mu)}{f_2(\mu)} - \frac{f_1'(\mu)}{f_1(\mu)} - \frac{f_2'(\lambda)}{f_2(\lambda)} + \frac{f_1'(\lambda)}{f_1(\lambda)}\right\} dS = 0;$$
Now the factor within { } cannot vanish for all values of λ and μ, unless the functions devoted by f_1 and f_2 be identical, or only differ by a numerical factor.

15. It will be well to transform the expression

$$\iint e\, V_1 V_2\, dS$$

to its equivalent, in terms of υ, υ'.

For this purpose we observe that if ds, ds' be elements of the two lines of curvature through any point of the ellipsoid, $dS = ds\, ds'$.

Now,

ds^2 is the value of $dx^2 + dy^2 + dz^2$ when ϵ and υ' are constant,

ds'^2 ... ... ϵ and υ ...

and
$$x^2 = \frac{(\epsilon + a^2)(\upsilon + a^2)(\upsilon' + a^2)}{(b^2 - a^2)(c^2 - a^2)};$$

therefore if ϵ and υ' do not vary,

$$\frac{2dx}{x} = \frac{d\upsilon}{\upsilon + a^2};$$

$$\therefore\ dx = \frac{1}{2}\frac{x}{a^2 + \upsilon}\, d\upsilon.$$

Similarly
$$dy = \frac{1}{2}\frac{y}{b^2 + \upsilon}\, d\upsilon, \quad dz = \frac{1}{2}\frac{z}{c^2 + \upsilon}\, d\upsilon;$$

$$\therefore\ ds^2 = dx^2 + dy^2 + dz^2 = \frac{1}{4}\left\{\frac{x^2}{(a^2+\upsilon)^2} + \frac{y^2}{(b^2+\upsilon)^2} + \frac{z^2}{(c^2+\upsilon)^2}\right\} d\upsilon^2.$$

Again, differentiating with respect to ω the expression obtained for $\dfrac{x^2}{a^2+\omega} + \dfrac{y^2}{b^2+\omega} + \dfrac{z^2}{c^2+\omega} - 1$, we get

$$\frac{x^2}{(a^2+\omega)^2} + \frac{y^2}{(b^2+\omega)^2} + \frac{z^2}{(c^2+\omega)^2} = \frac{(\upsilon-\omega)(\upsilon'-\omega)}{(a^2+\omega)(b^2+\omega)(c^2+\omega)}$$

$$+ \frac{(\upsilon'-\omega)(\epsilon-\omega)}{(a^2+\omega)(b^2+\omega)(c^2+\omega)} + \frac{(\epsilon-\omega)(\upsilon-\omega)(\upsilon'-\omega)}{(a^2+\omega)^2(b^2+\omega)(c^2+\omega)} + \ldots;$$

therefore, putting $\omega = v$,

$$\frac{x^2}{(a^2+v)^2}+\frac{y^2}{(b^2+v)^2}+\frac{z^2}{(c^2+v)^2}=\frac{(v'-v)(\epsilon-v)}{(a^2+v)(b^2+v)(c^2+v)};$$

$$\therefore ds^2=\frac{1}{4}\frac{(v'-v)(\epsilon-v)}{(a^2+v)(b^2+v)(c^2+v)}dv^2.$$

A similar expression holding for ds'^2 we get

$$dS^2=-\frac{1}{16}\frac{(v'-v)^2(\epsilon-v)(\epsilon-v')}{(a^2+v)(b^2+v)(c^2+v)(a^2+v')(b^2+v')(c^2+v')}dv^2dv'^2.$$

Again, $$\frac{1}{e^2}=\frac{x^2}{(a^2+\epsilon)^2}+\frac{y^2}{(b^2+\epsilon)^2}+\frac{z^2}{(c^2+\epsilon)^2}$$

$$=\frac{(\epsilon-v')(\epsilon-v)}{(a^2+\epsilon)(b^2+\epsilon)(c^2+\epsilon)},$$

writing ϵ for ω in the expression above;

$$\therefore e^2dS^2=-\frac{1}{16}\frac{(a^2+\epsilon)(b^2+\epsilon)(c^2+\epsilon)(v'-v)^2}{(a^2+v)(b^2+v)(c^2+v)(a^2+v')(b^2+v')(c^2+v')}dv^2dv'^2.$$

It has been shewn that, integrating all over the surface, the limits of v are $-c^2$ and $-b^2$, those of v', $-b^2$ and $-a^2$.

Hence, V_1, V_2, denoting two different ellipsoidal harmonics

$$\int_{-b^2}^{-c^2}\int_{-a^2}^{-b^2}\frac{V_1V_2(v'-v)\,dvdv'}{\{(a^2+v)(b^2+v)(c^2+v)(a^2+v')(b^2+v')(c^2+v')\}^{\frac{1}{2}}}=0.$$

The value of the expression $\iiint V^2dxdydz$, or its equivalent

$$abc\int_{-b^2}^{-c^2}\int_{-a^2}^{-b^2}\frac{V^2(v'-v)\,dvdv'}{\{(a^2+v)(b^2+v)(c^2+v)(a^2+v')(b^2+v')(c^2+v')\}^{\frac{1}{2}}},$$

in any particular case, is most conveniently obtained by expressing V as a function of x, y, z.

16. Before proceeding further with the discussion of ellipsoidal harmonics in general, we will consider the special case in which the ellipsoid is one of revolution. We must enquire what modification this will introduce in the quantities which we have denoted by α, β, γ, viz.

$$\alpha = \int_\epsilon^\infty \frac{d\psi}{(a^2+\psi)^{\frac{1}{2}}(b^2+\psi)^{\frac{1}{2}}(c^2+\psi)^{\frac{1}{2}}},$$

$$\beta = \int_\upsilon^{-c^2} \frac{d\psi}{(a^2+\psi)^{\frac{1}{2}}(b^2+\psi)^{\frac{1}{2}}(c^2+\psi)^{\frac{1}{2}}},$$

$$\gamma = \int_{-a^2}^{\upsilon'} \frac{d\psi}{(a^2+\psi)^{\frac{1}{2}}(b^2+\psi^2)^{\frac{1}{2}}(c^2+\psi)^{\frac{1}{2}}}$$

and in the differential equation

$$(\upsilon-\upsilon')\frac{d^2V}{d\alpha^2}+(\upsilon'-\epsilon)\frac{d^2V}{d\beta^2}+(\epsilon-\upsilon)\frac{d^2V}{d\gamma^2}=0.$$

We will first suppose the axis of revolution to be the greatest axis of the ellipsoid, which is equivalent to supposing $b^2=c^2$. To transform α and γ, put $a^2+\psi=\theta^2$, $a^2+\epsilon=\eta^2$, $a^2+\upsilon'=\omega^2$; we then obtain

$$\alpha = 2\int_\eta^\infty \frac{d\theta}{\theta^2-a^2+b^2} = \frac{1}{(a^2-b^2)^{\frac{1}{2}}}\log\frac{\eta+(a^2-b^2)^{\frac{1}{2}}}{\eta-(a^2-b^2)^{\frac{1}{2}}},$$

$$\gamma = 2\int_0^\omega \frac{d\theta}{\theta^2-a^2+b^2} = \frac{1}{(a^2-b^2)^{\frac{1}{2}}}\log\frac{(a^2-b^2)^{\frac{1}{2}}-\omega}{(a^2-b^2)^{\frac{1}{2}}+\omega}.$$

To transform β, we must proceed as follows.

Put $\psi=-c^2\cos^2\varpi-b^2\sin^2\varpi$, $\upsilon=-c^2\cos^2\phi-b^2\sin^2\phi$, we then get generally

$$b^2+\psi=(b^2-c^2)\cos^2\varpi,\quad c^2+\psi=(c^2-b^2)\sin^2\varpi;$$

$$d\psi=2(c^2-b^2)\cos\varpi\sin\varpi\, d\varpi;$$

$$\therefore\ \beta=\frac{2}{\sqrt{-1}}\int_\phi^0\frac{d\varpi}{(a^2-b^2)^{\frac{1}{2}}}=\frac{2\sqrt{-1}\,\phi}{(a^2-b^2)^{\frac{1}{2}}}.$$

Hence,
$$\frac{d}{d\alpha} = -\frac{1}{2}(\eta^2 - a^2 + b^2)\frac{d}{d\eta},$$

$$\frac{d}{d\gamma} = \frac{1}{2}(\omega^2 - a^2 + b^2)\frac{d}{d\omega},$$

$$\frac{d}{d\beta} = \frac{(a^2 - b^2)^{\frac{1}{2}}}{2\sqrt{-1}}\frac{d}{d\phi}.$$

Also, $\epsilon = \eta^2 - a^2$, $\upsilon' = \omega^2 - a^2$, $\upsilon = -b^2$, and our differential equation becomes

$$(a^2 - b^2 - \omega^2)\left\{(\eta^2 - a^2 + b^2)\frac{d}{d\eta}\right\}^2 V$$

$$+ (\eta^2 - a^2 + b^2)\left\{(\omega^2 - a^2 + b^2)\frac{d}{d\omega}\right\}^2 V$$

$$- (\omega^2 - \eta^2)(a^2 - b^2)\frac{d^2 V}{d\phi^2} = 0,$$

or
$$(\omega^2 - a^2 + b^2)\left\{(\eta^2 - a^2 + b^2)\frac{d}{d\eta}\right\}^2 V$$

$$- (\eta^2 - a^2 + b^2)\left\{(\omega^2 - a^2 + b^2)\frac{d}{d\omega}\right\}^2 V$$

$$- (a^2 - b^2)(\eta^2 - \omega^2)\frac{d^2 V}{d\phi^2} = 0.$$

This equation may be satisfied in the following ways.

First, in a manner altogether independent of ϕ, by supposing V to be the product of a function of η and the same function of ω, this function, which we will for the present denote by $f(\eta)$ or $f(\omega)$, being determined by the equation

$$\frac{d}{d\eta}\left\{(\eta^2 - a^2 + b^2)\frac{d}{d\eta}\right\} f(\eta) = mf(\eta),$$

or
$$\frac{d}{d\omega}\left\{(\omega^2 - a^2 + b^2)\frac{d}{d\omega}\right\} f(\omega) = mf(\omega).$$

Secondly, by supposing $\frac{d^2 V}{d\phi^2}$ a constant multiple of V, $= -\sigma^2 V$, suppose.

Our equation may then be written

$$(\omega^2 - a^2 + b^2)\left\{(\eta^2 - a^2 + b^2)\frac{d}{d\eta}\right\}^2 V$$

$$- (\eta^2 - a^2 + b^2)\left\{(\omega^2 - a^2 + b^2)\frac{d}{d\omega}\right\}^2 V$$

$$- \sigma^2 (a^2 - b^2)\{(\omega^2 - a^2 + b^2) - (\eta^2 - a^2 + b^2)\} V = 0,$$

which may be satisfied by supposing the factor of V independent of ϕ to be of the form $F(\eta)\, F(\omega)$, where

$$\left\{(\eta^2 - a^2 + b^2)\frac{d}{d\eta}\right\}^2 F(\eta) - \sigma^2 (a^2 - b^2)\, F(\eta) = m\,(\eta^2 - a^2 + b^2)\, F(\eta),$$

$$\left\{(\omega^2 - a^2 + b^2)\frac{d}{d\omega}\right\}^2 F(\omega) - \sigma^2 (a^2 - b^2)\, F(\omega) = m\,(\omega^2 - a^2 + b^2)\, F(\omega).$$

The factor involving ϕ will be of the form

$$A \cos \sigma\phi + B \sin \sigma\phi.$$

Now, returning to the equation

$$\frac{d}{d\eta}\left\{(\eta^2 - a^2 + b^2)\frac{d}{d\eta}\right\} f(\eta) = mf(\eta),$$

we see that, supposing the index of the highest power of η involved in $f(\eta)$ to be i, we must have $m = i(i+1)$.

Now, it will be observed that η may have any value however great, but that ω^2, which is equal to $a^2 + v'$, must lie between $a^2 - b^2$ and 0. Hence, putting $\omega^2 = (a^2 - b^2)\,\mu^2$, where μ^2 must lie between 0 and 1, we get

$$\frac{d}{d\mu}\left\{(1 - \mu^2)\frac{d}{d\mu}\right\} f\{(a^2 - b^2)^{\frac{1}{2}} \mu\} + i\,(i+1)\, f\{(a^2 - b^2)^{\frac{1}{2}} \mu\} = 0.$$

Hence this equation is satisfied by $f\{(a^2-b^2)^{\frac{1}{2}}\mu\} = CP_i$, C being a constant; and supposing $C=1$ we obtain the following series of values for $f(\omega)$,

$$i=0,\quad f(\omega)=1,$$

$$i=1,\quad f(\omega)=\frac{\omega}{(a^2-b^2)^{\frac{1}{2}}},$$

$$i=2,\quad f(\omega)=\frac{3\omega^2-(a^2-b^2)}{2(a^2-b^2)},$$

$$i=3,\quad f(\omega)=\frac{5\omega^3-3\omega(a^2-b^2)}{2(a^2-b^2)^{\frac{3}{2}}},$$

....

Exactly similar expressions may be obtained for $f(\eta)$, and these, when the attraction of ellipsoids is considered, will apply to all points within the ellipsoid. But they will be inadmissible for external points, since η is susceptible of indefinite increase.

The form of integral to be adopted in this case will be obtained by taking the other solution of the differential equation for the determination of $f(\eta)$, i.e. the zonal harmonic of the second kind, which is of the form $Q_i\left\{\frac{\eta}{(a^2-b^2)^{\frac{1}{2}}}\right\}$, where

$$Q_i\left\{\frac{\eta}{(a^2-b^2)^{\frac{1}{2}}}\right\}=P_i\left\{\frac{\eta}{(a^2-b^2)^{\frac{1}{2}}}\right\}\int_\eta^\infty\frac{d\theta}{\left[P_i\left\{\frac{\theta}{(a^2-b^2)^{\frac{1}{2}}}\right\}\right]^2(\theta^2-a^2+b^2)}.$$

Or, putting $\eta^2=(a^2-b^2)\nu^2$, $\theta^2=(a^2-b^2)\lambda^2$, we may write

$$Q_i(\nu)=P_i(\nu)\int_\nu^\infty\frac{d\lambda}{\{P_i(\lambda)\}^2(\lambda^2-1)}.$$

17. We may now consider what is the meaning of the quantities denoted by η and ω. They are the values of ϑ which satisfy the equation

$$\frac{x^2}{\vartheta^2}+\frac{y^2+z^2}{\vartheta^2-a^2+b^2}=1,$$

and are therefore the semi-axes of revolution of the surfaces confocal with the given ellipsoid, which pass through the point x, y, z. One of these surfaces is an ellipsoid, and its semi-axis is η. The other is an hyperboloid of two sheets whose semi-axis is ω.

Now, if θ be the eccentric angle of the point x, y, z, measured from the axis of revolution, we shall have

$$x^2 = \eta^2 \cos^2 \theta.$$

But also, since η^2, ω^2, are the two values of ϑ^2 which satisfy the equation of the surface,

$$\eta^2\omega^2 = (a^2 - b^2)\, x^2.$$

Hence $$\omega^2 = (a^2 - b^2) \cos^2 \theta,$$

and we have already put

$$\omega^2 = (a^2 - b^2)\, \mu^2,$$

whence the quantity which we have already denoted by μ is found to be the cosine of the eccentric angle of the point x, y, z considered with reference to the ellipsoid confocal with the given one, passing through the point x, y, z. We have thus a method of completely representing the potential of an ellipsoid of revolution for any distribution of density symmetrical about its axis, by means of the axis of revolution of the confocal ellipsoid passing through the point at which the potential is required, and the eccentric angle of the point with reference to the confocal ellipsoid. For any such distribution can be expressed, precisely as in the case of a sphere, by a series of zonal harmonic functions of the eccentric angle.

18. When the distribution is not symmetrical, we must have recourse to the form of solution which involves the factor $A \cos \sigma\phi + B \sin \sigma\phi$. It will be seen that, supposing F to represent a function of the degree i, and putting $m = i(i+1)$, the equation which determines $F(\omega)$ is of exactly the same form as that for a tesseral spherical harmonic. For $F(\eta)$, if the point be within the ellipsoid, we adopt the same form,

if without, representing the tesseral spherical harmonic by $T_i^{(\sigma)}\left\{\frac{\eta}{(a-b^2)^{\frac{1}{2}}}\right\}$, or $T_i^{(\sigma)}(\nu)$, we adopt the form

$$T_i^{(\sigma)}(\nu)\int_\nu^\infty \frac{d\lambda}{\overline{T_i^{(\sigma)}(\lambda)}|^2(\lambda^2-1)}.$$

19. It may be interesting to trace the connexion of spherical harmonics with the functions just considered. This may be effected by putting $b^2=a^2$. We see then that η will become equal to the radius of the concentric sphere passing through the point, and $\eta^2-a^2+b^2$ will become equal to η^2. Hence the equation for the determination of $f(\eta)$ will become

$$\frac{d}{d\eta}\left(\eta^2\frac{d}{d\eta}\right)f(\eta)=i(i+1)f(\eta),$$

which is satisfied by putting $f(\eta)=\eta^i$, or $\eta^{-(i+1)}$. The former solution is adapted to the case of an internal, the latter to that of an external point.

With regard to $f(\omega)$, it will be seen that the confocal hyperboloid becomes a cone, and therefore ω becomes indefinitely small. But μ, which is equal to $\frac{\omega}{(a^2-b^2)^{\frac{1}{2}}}$, remains finite, being in fact equal to $\frac{x}{\eta}$ or $\cos\theta$. Hence $f(\mu)$ becomes the zonal spherical harmonic.

Again, the tesseral equations, for the determination of $F(\eta)$, $F(\omega)$, become

$$\left(\eta^2\frac{d}{d\eta}\right)^2F(\eta)=i(i+1)\eta^2F(\eta),$$

which are satisfied by $F(\eta)=\eta^i$ or $\eta^{-(i+1)}$.

And, writing for ω^2, $(a^2-b^2)\mu^2$, we have, putting $F(\omega)=\chi(\mu)$,

$$\left\{(\mu^2-1)\frac{d}{d\mu}\right\}^2\chi(\mu)+\sigma^2\chi(\mu)=i(i+1)(\mu^2-1)\chi(\mu),$$

which gives $\chi(\mu)=T_i^{(\sigma)}(\mu)$.

20. We will next consider the case in which the axis of revolution is the least axis of the ellipsoid, which is equivalent to supposing $a^2 = b^2$. To transform α and β, put $c^2 + \psi = \theta^2$, $c^2 + \epsilon = \eta^2$, $c^2 + \upsilon = \omega^2$, we thus obtain

$$\alpha = 2\int_\eta^\infty \frac{d\theta}{a^2 - c^2 + \theta^2} = \frac{2}{(a^2 - c^2)^{\frac{1}{2}}} \tan^{-1} \frac{(a^2 - c^2)^{\frac{1}{2}}}{\eta},$$

$$\beta = 2\int_\omega^0 \frac{d\theta}{a^2 - c^2 + \theta^2} = -\frac{2}{(a^2 - c^2)^{\frac{1}{2}}} \tan^{-1} \frac{\omega}{(a^2 - c^2)^{\frac{1}{2}}}.$$

To transform γ, we must proceed as follows:

Put $\psi = -a^2 \sin^2 \varpi - b^2 \cos^2 \varpi$, $\upsilon' = -a^2 \sin^2 \phi - b^2 \cos^2 \phi$, we then get, generally,

$$a^2 + \psi = (a^2 - b^2)\cos^2\varpi, \quad b^2 + \psi = -(a^2 - b^2)\sin^2\varpi,$$

$$c^2 + \psi = c^2 - a^2 \sin^2\phi - b^2\cos^2\phi, \quad d\psi = -2(a^2 - b^2)\sin\varpi\cos\varpi\, d\varpi.$$

Hence

$$\gamma = 2\int_\phi^0 \frac{d\varpi}{(a^2\sin^2\varpi + b^2\cos^2\varpi - c^2)^{\frac{1}{2}}} = -\frac{2\phi}{(a^2 - c^2)^{\frac{1}{2}}} \text{ if } a^2 = b^2.$$

Hence,

$$\frac{d}{d\alpha} = -\frac{1}{2}(a^2 - c^2 + \eta^2)\frac{d}{d\eta},$$

$$\frac{d}{d\beta} = -\frac{1}{2}(a^2 - c^2 + \omega^2)\frac{d}{d\omega},$$

$$\frac{d}{d\gamma} = -\frac{1}{2}(a^2 - c^2)^{\frac{1}{2}}\frac{d}{d\phi};$$

also,

$$\epsilon = \eta^2 - c^2,$$

$$\upsilon = \omega^2 - c^2,$$

$$\upsilon' = -a^2,$$

and our differential equation becomes

$$(a^2 - c^2 + \omega^2)\left\{(a^2 - c^2 + \eta^2)\frac{d}{d\eta}\right\}^2 V$$

$$- (a^2 - c^2 + \eta^2)\left\{(a^2 - c^2 + \omega^2)\frac{d}{d\omega}\right\}^2 V$$

$$+ (\eta^2 - \omega^2)(a^2 - c^2)\frac{d^2 V}{d\phi^2} = 0.$$

We will first consider how this equation may be satisfied by values of V independent of ϕ.

We may then suppose V to be the product of a function of η, and the same function of ω, this function, which we will suppose to be of the degree i, being determined by the equation

$$\frac{d}{d\eta}\left\{(a^2-c^2+\eta^2)\frac{d}{d\eta}\right\}f(\eta)=i(i+1)f(\eta),$$

$$\frac{d}{d\omega}\left\{(a^2-c^2+\omega^2)\frac{d}{d\omega}\right\}f(\omega)=i(i+1)f(\omega).$$

On comparing this with the ordinary differential equation for a zonal harmonic, it will be seen that, on account of a^2 being greater than c^2, the signs of the several terms in the series for $f(\eta)$ will be all the same, instead of being alternately positive and negative. We shall thus have

$$i=1,\quad f(\eta)=\frac{\eta}{(a^2-c^2)^{\frac{1}{2}}},$$

$$i=2,\quad f(\eta)=\frac{3\eta^2+a^2-c^2}{2(a^2-c^2)},$$

$$i=3,\quad f(\eta)=\frac{5\eta^3+3(a^2-c^2)\eta}{2(a^2-c^2)^{\frac{3}{2}}},$$

$$i=4,\quad f(\eta)=\frac{35\eta^4+30(a^2-c^2)\eta^2+3(a^2-c^2)^2}{8(a^2-c^2)^2};$$

and generally

$$f(\eta)=\frac{1}{(a^2-c^2)^{\frac{i}{2}}}\frac{1}{2\,.\,4\,.\,6\ldots 2i}\frac{d^i}{d\eta^i}(\eta^2+a^2-c^2)^i.$$

We will denote the general value of $f(\eta)$ by $p_i\left\{\dfrac{\eta}{(a^2-c^2)^{\frac{1}{2}}}\right\}$, or, writing $\eta=(a^2-c^2)^{\frac{1}{2}}\nu$, by $p_i(\nu)$.

For external points, we must adopt for $f(\eta)$ a function which we will represent by $q_i\left\{\frac{\eta}{(a^2-c^2)^{\frac{1}{2}}}\right\}$, or $q_i(\nu)$, which will be equal to

$$p_i\left\{\frac{\eta}{(a^2-c^2)^{\frac{1}{2}}}\right\}\int_\eta^\infty \frac{d\theta}{\left[p_i\left\{\frac{\theta}{(a^2-c^2)^{\frac{1}{2}}}\right\}\right]^2(\theta^2+a^2-c^2)}.$$

It is clear that $f(\omega)$ may be expressed in exactly the same way. But it will be remembered that η^2 and ω^2 are the two values of ϑ^2 which satisfy the equation

$$\frac{x^2+y^2}{a^2-c^2+\vartheta^2}+\frac{z^2}{\vartheta^2}=1.$$

Hence η, as before, is the semi-axis of revolution of the confocal ellipsoid passing through the point (x, y, z). But $\eta^2\omega^2=-(a^2-c^2)z^2$, an essentially negative quantity, since a^2 is greater than c^2. Hence ω^2 is essentially negative. Now, if θ be the eccentric angle of the point (x, y, z) measured from the axis of revolution, we have $z^2=\eta^2\cos^2\theta$. Hence

$$\eta^2\omega^2=-(a^2-c^2)\eta^2\cos^2\theta,$$

and therefore
$$\omega^2=-(a^2-c^2)\cos^2\theta$$
$$=-(a^2-c^2)\mu^2, \text{ suppose.}$$

Hence the equation for the determination of $f(\omega)$ assumes the form

$$\frac{d}{d\mu}\left\{(1-\mu^2)\frac{d}{d\mu}\right\}f(\omega)+i(i+1)f(\omega)=0,$$

the ordinary equation for a zonal spherical harmonic. Hence we may write

$$f(\omega)=P_i(\mu),$$

μ being the cosine of the eccentric angle of the point x, y, z, considered with reference to the confocal ellipsoid passing through it.

21. We have thus discussed the form of the potential, corresponding to a distribution of attracting matter, symmetrical about the axis. When the distribution is not symmetrical, but involves ϕ in the form $A\cos\sigma\phi + B\sin\sigma\phi$, we replace, as before, $P_i(\mu)$ by $T_i^{(\sigma)}(\mu)$, and $p_i(\mu)$ by a function $t_i^{(\sigma)}(\nu)$ determined by the equation

$$t_i^{(\sigma)}(\nu) = (1+\nu^2)^{\frac{\sigma}{2}} \frac{d^\sigma}{d\nu^\sigma} p_i(\nu),$$

and $q_i(\nu)$ by
$$t_i^{(\sigma)}(\nu) \int_\nu^\infty \frac{d\lambda}{\overline{t_i^{(\sigma)}(\lambda)}|^2 (\lambda^2+1)}.$$

22. As an application of these formulæ, consider the following question.

Attracting matter is distributed over the shell whose surface is represented by the equation $\frac{x^2}{a^2} + \frac{y^2+z^2}{b^2} = 1$, so that its volume density at any point is $P_i(\mu)$, μ being the cosine of the eccentric angle, measured from the axis of revolution; required to determine the potential at any point, external or internal.

The potential at any internal point will be of the form

$$CP_i(\mu)\,P_i(\nu) \ldots\ldots\ldots\ldots\ldots\ldots (1),$$

and at an external point, of the form

$$C'P_i(\mu)\,Q_i(\nu) \ldots\ldots\ldots\ldots\ldots\ldots (2),$$

where $(a^2-b^2)^{\frac{1}{2}}\nu =$ the semi-axis of the figure of the confocal ellipsoid of revolution passing through the point (μ, ν).

Now the expressions (1) and (2) must be equal at the surface of the ellipsoid, where $\nu = \frac{a}{(a^2-b^2)^{\frac{1}{2}}}$.

Hence

$$CP_i(\mu)\,P_i\left\{\frac{a}{(a^2-b^2)^{\frac{1}{2}}}\right\} = C'P_i(\mu)\,Q_i\left\{\frac{a}{(a^2-b^2)^{\frac{1}{2}}}\right\}.$$

But generally

$$Q_i(\nu) = P_i(\nu) \int_\nu^\infty \frac{d\lambda}{\overline{P_i(\lambda)}|^2 (\lambda^2 - 1)}.$$

Hence

$$Q_i\left\{\frac{a}{(a^2-b^2)^{\frac{1}{2}}}\right\} = P_i\left\{\frac{a}{(a^2-b^2)^{\frac{1}{2}}}\right\} \int_{\frac{a}{(a^2-b^2)^{\frac{1}{2}}}}^\infty \frac{d\lambda}{\overline{P_i(\lambda)}|^2 (\lambda^2 - 1)};$$

$$\therefore\ CP_i\left\{\frac{a}{(a^2-b^2)^{\frac{1}{2}}}\right\} = C'P_i\left\{\frac{a}{(a^2-b^2)^{\frac{1}{2}}}\right\} \int_{\frac{a}{(a^2-b^2)^{\frac{1}{2}}}}^\infty \frac{d\lambda}{\overline{P_i(\lambda)}|^2 (\lambda^2 - 1)}.$$

We may therefore, putting $C' = AP_i\left\{\frac{a}{(a^2-b^2)^{\frac{1}{2}}}\right\}$, write

$$C = AQ_i\left\{\frac{a}{(a^2-b^2)^{\frac{1}{2}}}\right\},$$

and we thus express the potentials as follows:

$$AP_i(\mu)\,P_i(\nu)\,Q_i\left\{\frac{a}{(a^2-b^2)^{\frac{1}{2}}}\right\} \text{ at an internal point,}$$

$$AP_i(\mu)\,Q_i(\nu)\,P_i\left\{\frac{a}{(a^2-b^2)^{\frac{1}{2}}}\right\} \text{ at an external point.}$$

Or, substituting for Q_i its value in terms of P_i,

$$V_1 = AP_i(\mu)\,P_i(\nu)\,P_i\left\{\frac{a}{(a^2-b^2)^{\frac{1}{2}}}\right\} \int_{\frac{a}{(a^2-b^2)^{\frac{1}{2}}}}^\infty \frac{d\lambda}{\overline{P_i(\lambda)}|^2 (\lambda^2 - 1)}$$

at an internal point,

$$V_2 = AP_i(\mu)\,P_i(\nu)\,P_i\left\{\frac{a}{(a^2-b^2)^{\frac{1}{2}}}\right\} \int_\nu^\infty \frac{d\lambda}{\overline{P_i(\lambda)}|^2 (\lambda^2 - 1)}$$

at an external point.

Now, to determine A, we have, δa being the thickness of the shell at the extremity of the axis of revolution,

$$\rho = \frac{1}{4\pi} \frac{a}{\delta a \,.\, \eta} \left(\frac{dV_1}{d\eta} - \frac{dV_2}{d\eta}\right)_{\eta=a}$$

$$= \frac{1}{4\pi} \frac{a}{\delta a} \frac{1}{a^2-b^2} \left(\frac{dV_1}{d\nu} - \frac{dV_2}{d\nu}\right)_{\nu=\frac{a}{(a^2-b^2)^{\frac{1}{2}}}}$$

$$= \frac{1}{4\pi} \frac{a}{\delta a} \frac{1}{a^2-b^2} AP_i(\mu)\, P_i\left\{\frac{a}{(a^2-b^2)^{\frac{1}{2}}}\right\} P_i\left\{\frac{a}{(a^2-b^2)^{\frac{1}{2}}}\right\} \frac{1}{\left[P_i\left\{\frac{a}{(a^2-b^2)^{\frac{1}{2}}}\right\}\right]^2 \left(\frac{a^2}{a^2-b^2}-1\right)}$$

$$= \frac{1}{4\pi} \frac{a}{\delta a} \frac{A}{a^2-b^2} P_i(\mu) \frac{1}{\frac{a^2}{a^2-b^2}-1}$$

$$= \frac{1}{4\pi} \frac{a}{b^2\delta a} AP_i(\mu).$$

Hence, if $\rho = P_i(\mu)$, we obtain

$$A = 4\pi \frac{b^2 \delta a}{a} = 4\pi b \delta b.$$

And we thus obtain

$$V_1 = 4\pi\, b\delta b\, P_i(\mu)\, P_i(\nu)\, P_i\left\{\frac{a}{(a^2-b^2)^{\frac{1}{2}}}\right\} \int_{\frac{a}{(a^2-b^2)^{\frac{1}{2}}}}^{\infty} \frac{d\lambda}{\overline{P_i(\lambda)}|^2 (\lambda^2-1)}$$

$$= 4\pi b\delta b\, P_i(\mu)\, P_i(\nu)\, Q_i\left\{\frac{a}{(a^2-b^2)^{\frac{1}{2}}}\right\}.$$

$$V_2 = 4\pi b\delta b\, P_i(\mu)\, P_i\left\{\frac{a}{(a^2-b^2)^{\frac{1}{2}}}\right\} P_i(\nu) \int_{\nu}^{\infty} \frac{d\lambda}{\overline{P_i(\lambda)}|^2 (\lambda^2-1)}$$

$$= 4\pi b\delta b\, P_i(\mu)\, Q_i(\nu)\, P_i\left\{\frac{a}{(a^2-b^2)^{\frac{1}{2}}}\right\}.$$

If the shell be represented by the equation

$$\frac{x^2+y^2}{a^2} + \frac{z}{c^2} = 1,$$

it may be shewn in a similar manner that we shall have

$$V_1 = 4\pi ada\, P_i(\mu)\, p_i(\nu)\, q_i \left\{\frac{c}{(a^2 - c^2)^{\frac{1}{2}}}\right\},$$

$$V_2 = 4\pi ada\, P_i(\mu)\, q_i(\nu)\, p_i \left\{\frac{c}{(a^2 - c^2)^{\frac{1}{2}}}\right\}.$$

23. We may apply this result to the discussion of the following problem.

If the potential of a shell in the form of an ellipsoid of revolution about the greatest point be inversely proportional to the distance from one focus, find the potential at any internal point, and the density.

If the potential at P be inversely proportional to the distance from one focus S, and H be the other focus, we have,

$$HP + SP = 2\eta, \quad HP - SP = 2\omega,$$

$$\therefore\ SP = \eta - \omega.$$

Hence if M be the mass of the shell, V_2 the potential at any external point,

$$V_2 = \frac{M}{\eta - \omega}$$

$$= \frac{M}{(a^2 - b^2)^{\frac{1}{2}}} \frac{1}{\nu - \mu}$$

$$= \frac{M}{(a^2 - b^2)^{\frac{1}{2}}} \Sigma\,(2i + 1)\, P_i(\mu)\, Q_i(\nu).$$

Now, by what has just been seen, the internal potential, corresponding to $P_i(\mu)\, Q_i(\nu)$, is

$$P_i(\mu)\, P_i(\nu) \frac{Q_i\left\{\dfrac{a}{(a^2 - b^2)^{\frac{1}{2}}}\right\}}{P_i\left\{\dfrac{a}{(a^2 - b^2)^{\frac{1}{2}}}\right\}}.$$

Hence, if V_2 be the potential at any internal point,

$$V_2 = \frac{M}{(a^2-b^2)^{\frac{1}{2}}} \Sigma\,(2i+1) \frac{Q_i\left\{\dfrac{a}{(a^2-b^2)^{\frac{1}{2}}}\right\}}{P_i\left\{\dfrac{a}{(a^2-b^2)^{\frac{1}{2}}}\right\}} P_i(\mu)\,P_i(\nu).$$

And the volume density corresponding to $P_i(\mu)\;Q_i(\nu)$ is

$$\frac{P_i(\mu)}{4\pi b\delta b\,P_i\left\{\dfrac{a}{(a^2-b^2)^{\frac{1}{2}}}\right\}}.$$

Hence the density corresponding to the present distribution is

$$\rho = \frac{M}{4\pi\,(a^2-b^2)^{\frac{1}{2}}\,b\delta b} \Sigma\,(2i+1) \frac{P_i(\mu)}{P_i\left\{\dfrac{a}{(a^2-b^2)^{\frac{1}{2}}}\right\}}.$$

If V_2 had varied inversely as HP, we should have had

$$V_2 = \frac{M}{\eta+\omega},$$

and our results would have been obtained from the foregoing by changing the sign of ω, and therefore of μ.

24. Now, by adding these results together, we obtain the distributions of density, and internal potential, corresponding to

$$V_2 = \frac{M}{\eta-\omega} + \frac{M}{\eta+\omega} = M\frac{2\eta}{\eta^2-\omega^2},$$

or, in geometrical language,

$$V_2 = \frac{M}{SP} + \frac{M}{HP} = M\frac{SP+HP}{SP\,.\,HP},$$

$= M$ multiplied by the axis of revolution of the confocal ellipsoid, and divided by the square on the conjugate semidiameter. We may express this by saying that the potential at any point on the ellipsoid is inversely proportional to the square on the conjugate semi-diameter, or directly as the square on the perpendicular on the tangent plane.

Corresponding to this, we shall have, writing $2k$ for i, since only even values of i will be retained,

$$V_1 = \frac{2M}{(a^2-b^2)^{\frac{1}{2}}} \Sigma (4k+1) \frac{Q_{2k}\left\{\frac{a}{(a^2-b^2)^{\frac{1}{2}}}\right\}}{P_{2k}\left\{\frac{a}{(a^2-b^2)^{\frac{1}{2}}}\right\}} P_{2k}(\mu) P_{2k}(\nu),$$

$$\rho = \frac{2M}{4\pi (a^2-b^2)^{\frac{1}{2}} b\delta b} \Sigma (4k+1) \frac{P_{2k}(\mu)}{P_{2k}\left\{\frac{a}{(a^2-b^2)^{\frac{1}{2}}}\right\}},$$

k being 0, or any positive integer.

Again, subtracting these results we get

$$V_2 = \frac{M}{\eta-\omega} - \frac{M}{\eta+\omega} = M\frac{2\omega}{\eta^2-\omega^2},$$

$=M$ multiplied by the distance from the equatoreal plane, and divided by the square on the conjugate semi-diameter.

This gives, writing $2k+1$ for i,

$$V_1 = \frac{2M}{(a^2-b^2)^{\frac{1}{2}}} \Sigma (4k+3) \frac{Q_{2k+1}\left\{\frac{a}{(a^2-b^2)^{\frac{1}{2}}}\right\}}{P_{2k+1}\left\{\frac{a}{(a^2-b^2)^{\frac{1}{2}}}\right\}} P_{2k+1}(\mu) P_{2k+1}(\nu),$$

$$\rho = \frac{2M}{4\pi (a^2-b^2)^{\frac{1}{2}} b\delta b} \Sigma (4k+3) \frac{P_{2k+1}(\mu)}{P_{2k+1}\left\{\frac{a}{(a^2-b^2)^{\frac{1}{2}}}\right\}}.$$

25. In attempting to discuss the problem analogous to this for an ellipsoid of revolution about its least axis, we see that since its foci are imaginary, the first problem would represent no real distribution. But if we suppose the external potential to be the *sum* or *difference* of two expressions, each inversely proportional to the distance from one focus, we

obtain a real distribution of potential—in the first case inversely proportional to the square on the conjugate semi-diameter, in the latter varying as the quotient of the distance from the equatoreal plane by the square on the conjugate semi-diameter.

It will be found, by a process exactly similar to that just adopted, that the distributions of internal potential, and density, respectively corresponding to these will be:

In the first case

$$V_1 = \frac{2M}{(a^2 - c^2)^{\frac{1}{2}}} \Sigma (4k+1) \frac{q_{2k}\left\{\dfrac{c}{(a^2-c^2)^{\frac{1}{2}}}\right\}}{p_{2k}\left\{\dfrac{c}{(a^2-c^2)^{\frac{1}{2}}}\right\}} P_{2k}(\mu)\, p_{2k}(\nu),$$

$$\rho = \frac{2M}{4\pi (a^2 - c^2)^{\frac{1}{2}}\, a\delta a} \Sigma (4k+1) \frac{P_{2k}(\mu)}{p_{2k}\left\{\dfrac{c}{(a^2-c^2)^{\frac{1}{2}}}\right\}},$$

k being 0, or any positive integer.

In the second case

$$V_1 = \frac{2M}{(a^2 - c^2)^{\frac{1}{2}}} \Sigma (4k+3) \frac{q_{2k+1}\left\{\dfrac{c}{(a^2-c^2)^{\frac{1}{2}}}\right\}}{p_{2k+1}\left\{\dfrac{c}{(a^2-c^2)^{\frac{1}{2}}}\right\}} P_{2k+1}(\mu)\, p_{2k+1}(\nu),$$

$$\rho = \frac{2M}{4\pi (a^2 - c^2)^{\frac{1}{2}}\, a\delta a} \Sigma (4k+3) \frac{P_{2k+1}(\mu)}{p_{2k+1}\left\{\dfrac{c}{(a^2-c^2)^{\frac{1}{2}}}\right\}},$$

k being 0, or any positive integer.

26. We may now resume the consideration of the ellipsoid with three unequal axes, and may shew how, when the potential at every point of the surface of an ellipsoidal shell is known, the functions which we are considering may be employed to determine its value at any internal or external point. We will begin by considering some special cases,

by which the general principles of the method may be made more intelligible.

27. First, suppose that the potential at every point of the surface of the ellipsoid is proportional to $x = \frac{V_0 x}{a}$ suppose.

In this case, since x when substituted for V, satisfies the equation $\nabla^2 V = 0$, we see that $V_0 \frac{x}{a}$ will also be the potential at any internal point. But this value will not be admissible at external points, since x becomes infinite at an infinite distance.

Now, transforming to elliptic co-ordinates

$$x = \left\{\frac{(\epsilon + a^2)(\upsilon + a^2)(\upsilon' + a^2)}{(a^2 - b^2)(a^2 - c^2)}\right\}^{\frac{1}{2}}.$$

And the expression

$$\frac{V_0}{a}\left\{\frac{(\epsilon+a^2)(\upsilon+a^2)(\upsilon'+a^2)}{(a^2-b^2)(a^2-c^2)}\right\}^{\frac{1}{2}} \int_\epsilon^\infty \frac{d\psi}{(\psi+a^2)\{(\psi+a^2)(\psi+b^2)(\psi+c^2)\}^{\frac{1}{2}}}$$
$$\div \int_0^\infty \frac{d\psi}{(\psi+a^2)\{(\psi+a^2)(\psi+b^2)(\psi+c^2)\}^{\frac{1}{2}}}$$

satisfies, as has already been seen, the equation $\nabla^2 V = 0$, is equal to $V_0 \frac{x}{a}$ at the surface of the ellipsoid, and vanishes at an infinite distance. This is therefore the value of the potential at any external point. It may of course be written

$$\frac{V_0 x}{a} \int_\epsilon^\infty \frac{d\psi}{(\psi+a^2)\{(\psi+a^2)(\psi+b^2)(\psi+c^2)\}^{\frac{1}{2}}}$$
$$\div \int_0^\infty \frac{d\psi}{(\psi+a^2)\{(\psi+a^2)(\psi+b^2)(\psi+c^2)\}^{\frac{1}{2}}}.$$

28. Next, suppose that the potential at every point of the surface is proportional to $yz = V_0 \frac{yz}{bc}$, suppose. In this

case, as in the last, we see that, since yz when substituted for V, satisfies the eqnation $\nabla^2 V = 0$, the potential at any internal point will be $V_0 \frac{yz}{bc}$; while, substituting for y, z their values in terms of elliptic co-ordinates we obtain for the potential at any external point

$$\frac{V_0 yz}{bc} \int_\epsilon^\infty \frac{d\psi}{(\psi + b^2)(\psi + c^2)\{(\psi + a^2)(\psi + b^2)(\psi + c^2)\}^{\frac{1}{2}}}$$

$$\div \int_0^\infty \frac{d\psi}{(\psi + b^2)(\psi + c^2)\{(\psi + a^2)(\psi + b^2).(\psi + c^2)\}^{\frac{1}{2}}}.$$

29. We will next consider the case in which the potential, at every point of the surface, varies as $x^2 = V_0 \frac{x^2}{a^2}$ suppose. This case materially differs from the two just considered, for since x^2 does not, when substituted for V, satisfy the equation $\nabla^2 V = 0$, the potential at internal points cannot in general be proportional to x^2. We have therefore first to investigate a function of x, y, z, or of ϵ, υ, υ' which shall satisfy the equation $\nabla^2 V = 0$, shall not become infinite within the surface of the ellipsoid, and shall be equal to x^2 on its surface.

Now we know that, generally

$$(b^2 + \omega)(c^2 + \omega) x^2 + (c^2 + \omega)(a^2 + \omega) y^2 + (a^2 + \omega)(b^2 + \omega) z^2$$
$$- (a^2 + \omega)(b^2 + \omega)(c^2 + \omega) = (\epsilon - \omega)(\upsilon - \omega)(\upsilon' - \omega).$$

And, if θ_1, θ_2 be the two values of ω which satisfy the equation

$$(b^2 + \omega)(c^2 + \omega) + (c^2 + \omega)(a^2 + \omega) + (a^2 + \omega)(b^2 + \omega) = 0 \ldots (1),$$

we see that

$$\nabla^2 (\epsilon - \theta_1)(\upsilon - \theta_1)(\upsilon' - \theta_1) = 0,$$

and

$$\nabla^2 (\epsilon - \theta_2)(\upsilon - \theta_2)(\upsilon' - \theta_2) = 0.$$

And, by properly determining the coefficients A_0, A_1, A_2, it is possible to make

$$A_0 + A_1 (\epsilon - \theta_1)(\upsilon - \theta_1)(\upsilon' - \theta_1) + A_2 (\epsilon - \theta_2)(\upsilon - \theta_2)(\upsilon' - \theta_2) \ldots (2)$$

$$= \frac{V_0 x^2}{a^2} \text{ when } b^2 c^2 x^2 + c^2 a^2 y^2 + a^2 b^2 z^2 - a^2 b^2 c^2 = 0.$$

Hence, the expression (2) when A_0, A_1, A_2 are properly determined will satisfy all the necessary conditions for an internal potential, and will therefore be the potential for every internal point.

Now, we have in general

$$(b^2+\theta_1)(c^2+\theta_1)x^2+(c^2+\theta_1)(a^2+\theta_1)y^2+(a^2+\theta_1)^2(b^2+\theta_1)z^2$$
$$-(a^2+\theta_1)(b^2+\theta_1)(c^2+\theta_1)=(\epsilon-\theta_1)(\upsilon-\theta_1)(\upsilon'-\theta_1)$$
$$(b^2+\theta_2)(c^2+\theta_2)x^2+(c^2+\theta_2)(a^2+\theta_2)y^2+(a^2+\theta_2)(b^2+\theta_2)z^2$$
$$-(a^2+\theta_2)(b^2+\theta_2)(c^2+\theta_2)=(\epsilon-\theta_2)(\upsilon-\theta_2(\upsilon'-\theta_2)$$

and, over the surface

$$b^2c^2x^2+c^2a^2y^2+a^2b^2z^2-a^2b^2c^2=0.$$

Hence, ϑ being any quantity whatever, we have, all over the surface,

$$(b^2+\vartheta)(c^2+\vartheta)x^2+(c^2+\vartheta)(a^2+\vartheta)y^2+(a^2+\vartheta)(b^2+\vartheta)z^2$$
$$-(a^2+\vartheta)(b^2+\vartheta)(c^2+\vartheta)$$
$$=\frac{\vartheta(\vartheta-\theta_2)}{\theta_1(\theta_1-\theta_2)}(\epsilon-\theta_1)(\upsilon-\theta_1)(\upsilon'-\theta_1)$$
$$+\frac{\vartheta(\vartheta-\theta_1)}{\theta_2(\theta_2-\theta_1)}(\epsilon-\theta_2)(\upsilon-\theta_2)(\upsilon'-\theta_2)-\vartheta(\vartheta-\theta_1)(\vartheta-\theta_2),$$

and therefore, putting $\vartheta=-a^2$,

$$(a^2-b^2)(a^2-c^2)x^2=\frac{a^2(a^2+\theta_2)}{\theta_1(\theta_1-\theta_2)}(\epsilon-\theta_1)(\upsilon-\theta_1)(\upsilon'-\theta_1)$$
$$+\frac{a^2(a^2+\theta_1)}{\theta_2(\theta_2-\theta_1)}(\epsilon-\theta_2)(\upsilon-\theta_2)(\upsilon'-\theta_2)+a^2(a^2+\theta_1)(a^2+\theta_2).$$

Hence, the right-hand member of this equation possesses all the necessary properties of an internal potential. It satisfies the general differential equation of the second order, does not become infinite *within* the shell, and is proportional to x^2 all over the surface.

We observe, by equation (1), that

$$(b^2+\omega)(c^2+\omega)+(c^2+\omega)(a^2+\omega)+(a^2+\omega)(b^2+\omega)=3(\theta_1-\omega)(\theta_2-\omega)$$

identically, and therefore, writing $-a^2$ for ω,

$$(a^2-b^2)(a^2-c^2)=3(a^2+\theta_1)(a^2+\theta_2).$$

Hence, over the surface of the shell,

$$x^2=\frac{a^2}{3\theta_1(\theta_1-\theta_2)(a^2+\theta_1)}(\epsilon-\theta_1)(\upsilon-\theta_1)(\upsilon'-\theta_1)$$
$$+\frac{a^2}{3\theta_2(\theta_2-\theta_1)(a^2+\theta_2)}(\epsilon-\theta_2)(\upsilon-\theta_2)(\upsilon'-\theta_2)+\frac{a^2}{3},$$

and we therefore have, for the internal potential,

$$V_1=\frac{V_0}{3}\left\{\frac{(\epsilon-\theta_1)(\upsilon-\theta_1)(\upsilon'-\theta_1)}{\theta_1(\theta_1-\theta_2)(a^2+\theta_1)}+\frac{(\epsilon-\theta_2)(\upsilon-\theta_2)(\upsilon'-\theta_2)}{\theta_2(\theta_2-\theta_1)(a^2+\theta_2)}+1\right\}.$$

This is not admissible for external points, as it becomes infinite at an infinite distance. We must therefore substitute for the factor $\epsilon-\theta_1$

$$(\epsilon-\theta_1)\int_\epsilon^\infty\frac{d\psi}{(\psi-\theta_1)^2\{(\psi+a^2)(\psi+b^2)(\psi+c^2)\}^{\frac{1}{2}}}$$
$$\div\int_0^\infty\frac{d\psi}{(\psi-\theta_1)^2\{(\psi+a^2)(\psi+b^2)(\psi+c^2)\}^{\frac{1}{2}}},$$

with a similar substitution for $\epsilon-\theta_2$, thus giving, for the external potential,

$$V_2=\frac{V_0}{3}\left[\frac{(\epsilon-\theta_1)(\upsilon-\theta_1)(\upsilon'-\theta_1)}{\theta_1(\theta_1-\theta_2)(a^2+\theta_1)}\int_\epsilon^\infty\frac{d\psi}{(\psi-\theta_1)^2\{(\psi+a^2)(\psi+b^2)(\psi+c^2)\}^{\frac{1}{2}}}\right.$$
$$\div\int_0^\infty\frac{d\psi}{(\psi-\theta_1)^2\{(\psi+a^2)(\psi+b^2)(\psi+c^2)\}^{\frac{1}{2}}}$$
$$+\frac{(\epsilon-\theta_2)(\upsilon-\theta_2)(\upsilon'-\theta_2)}{\theta_2(\theta_2-\theta_1)(a^2+\theta_2)}\int_\epsilon^\infty\frac{d\psi}{(\psi-\theta_2)^2\{(\psi+a^2)(\psi+b^2)(\psi+c^2)\}^{\frac{1}{2}}}$$
$$\div\int_0^\infty\frac{d\psi}{(\psi-\theta_2)^2\{(\psi+a^2)(\psi+b^2)(\psi+c^2)\}^{\frac{1}{2}}}$$
$$\left.+\int_\epsilon^\infty\frac{d\psi}{\{(\psi+a^2)(\psi+b^2)(\psi+c^2)\}^{\frac{1}{2}}}\div\int_0^\infty\frac{d\psi}{\{\psi+a^2)(\psi+b^2)(\psi+c^2)\}^{\frac{1}{2}}}\right].$$

The distribution of density over the surface, corresponding to this distribution of potential, may be investigated by means of the formula

$$\rho = \frac{1}{2\pi}\frac{a}{da}\left(\frac{dV_1}{d\epsilon} - \frac{dV_2}{d\epsilon}\right)_{\epsilon=0},$$

or its equivalent in Art. 13 of this Chapter. We thus find that

$$\rho = \frac{1}{2\pi}\frac{a}{da}\frac{V_0}{3abc}\Bigg[-\frac{(\upsilon-\theta_1)(\upsilon'-\theta_1)}{\theta_1^2(\theta_1-\theta_2)(a^2+\theta_2)}$$

$$\div\int_0^\infty \frac{d\psi}{(\psi-\theta_1)^2\{(\psi+a^2)(\psi+b^2)(\psi+c^2)\}^{\frac{1}{2}}}$$

$$-\frac{(\upsilon-\theta_2)(\upsilon'-\theta_2)}{\theta_2^2(\theta_2-\theta_1)(a^2+\theta_2)}\div\int_0^\infty \frac{d\psi}{(\psi-\theta_2)^2\{(\psi+a^2)(\psi+b^2)(\psi+c^2)\}^{\frac{1}{2}}}$$

$$+1\div\int_0^\infty \frac{d\psi}{\{(\psi+a^2)(\psi+b^2)(\psi+c^2)\}^{\frac{1}{2}}}\Bigg].$$

30. The investigation just given, of the potential at an external point of a distribution of matter giving rise to a potential proportional to x^2 all over the surface, has an interesting practical application. For the Earth may be regarded as an ellipsoid of equilibrium (not necessarily with two of its axes equal) under the action of the mutual gravitation of its parts and of the centrifugal force. If, then, V denote the potential of the Earth at any point on or without its surface, and Ω the angular velocity of the Earth's rotation, we have, as the equation of its surface, regarded as a surface of equal pressure,

$$\left(\frac{dV}{dx} + \Omega^2 x\right)dx + \left(\frac{dV}{dy} + \Omega^2 y\right)dy + \frac{dV}{dz}dz = 0.$$

$$\therefore V + \frac{1}{2}\Omega^2(x^2+y^2) = \text{a constant, } \Pi \text{ suppose.}$$

Hence, if a, b, c denote the semi-axes of the Earth, we have, for the determination of V, the following conditions:

$$\frac{d^2V}{dx^2}+\frac{d^2V}{dy^2}+\frac{d^2V}{dz^2}=0 \quad\text{..................(1)},$$

$$V=0 \text{ at an infinite distance..............(2)},$$

$V=\Pi-\frac{1}{2}\Omega^2(x^2+y^2)$ when

$$\frac{x^2}{a^2}+\frac{y^2}{b^2}+\frac{z^2}{c^2}=1 \quad\text{.....................(3)}.$$

The term Π will, as we know, give rise to an external potential represented by

$$\Pi\int_\epsilon^\infty \frac{d\psi}{\{(\psi+a^2)(\psi+b^2)(\psi+c^2)\}^{\frac{1}{2}}} \div \int_0^\infty \frac{d\psi}{\{(\psi+a^2)(\psi+b^2)(\psi+c^2)\}^{\frac{1}{2}}}.$$

The two terms $-\frac{1}{2}\Omega^2x^2$, $-\frac{1}{2}\Omega^2y^2$, will give rise to terms which may be deduced from the value of V_2 just given by successively writing for V_0, $-\frac{1}{2}\Omega^2a^2$, and $-\frac{1}{2}\Omega^2b^2$, and (in the latter case) putting b^2 for a^2 throughout. We thus get

$$V=\left\{\Pi-\frac{1}{6}\Omega^2(a^2+b^2)\right\}\int_\epsilon^\infty \frac{d\psi}{\{(\psi+a^2)(\psi+b^2)(\psi+c^2)\}^{\frac{1}{2}}}$$

$$\div\int_0^\infty \frac{d\psi}{\{(\psi+a^2)(\psi+b^2)(\psi+c^2)\}^{\frac{1}{2}}}-\frac{\Omega^2}{6}\left(\frac{a^2}{a^2+\theta_1}+\frac{b^2}{b^2+\theta_1}\right)$$

$$\frac{(\epsilon-\theta_1)(v-\theta_1)(v'-\theta_1)}{\rho_1(\theta_1-\theta_2)}\int_\epsilon^\infty \frac{d\psi}{(\psi-\theta_1)^2\{(\psi+a^2)(\psi+b^2)(\psi+c^2)\}^{\frac{1}{2}}}$$

$$\div\int_0^\infty \frac{d\psi}{(\psi-\theta_1)^2\{(\psi+a^2)(\psi+b^2)(\psi+c^2)\}^{\frac{1}{2}}}-\frac{\Omega^2}{6}\left(\frac{a^2}{a^2+\theta_2}+\frac{b^2}{b^2+\theta_2}\right)$$

$$\frac{(\epsilon-\theta_2)(v-\theta_2)(v'-\theta_2)}{\theta_2(\theta_2-\theta_1)}\int_\epsilon^\infty \frac{d\psi}{(\psi-\theta_2)^2\{(\psi+a^2)(\psi+b^2)(\psi+c^2)\}^{\frac{1}{2}}}$$

$$\div\int_0^\infty \frac{d\psi}{(\psi-\theta_2)^2\{(\psi+a^2)(\psi+b^2)(\psi+c^2)\}^{\frac{1}{2}}}.$$

31. Any rational integral function V of x, y, z, which satisfies the equation $\nabla^2 V = 0$, can be expressed in a series of Ellipsoidal Harmonics of the degrees $0, 1, 2 \ldots i$ in x, y, z. For if V be of the degree i, the number of terms in V will be $\frac{(i+1)(i+2)(i+3)}{6}$. Now the condition $\nabla^2 V = 0$ is equivalent to the condition that a certain function of x, y, z of the degree $i-2$, vanishes identically, and this gives rise to $\frac{(i-1)\,i\,(i+1)}{6}$ conditions. Hence the number of independent constants in V is

$$\frac{(i+1)(i+2)(i+3)}{6} - \frac{(i-1)\,i\,(i+1)}{6},$$

or $(i+1)^2$. And the number of ellipsoidal harmonics of the degrees $0, 1, 2 \ldots i$ in x, y, z or of the degrees $0, \frac{1}{2}, 1, \frac{3}{2} \ldots \frac{i}{2}$ in ϵ, v, v', is, as shewn in Arts. 6 to 10 of this Chapter,

$$1 + 3 + 5 + \ldots + 2i + 1,$$

or $(i+1)^2$. Hence all the necessary conditions can be satisfied.

32. Again, suppose that attracting matter is distributed over the surface of an ellipsoidal shell according to a law of density expressed by any rational integral function of the co-ordinates. Let the dimensions of the highest term in this expression be i, then by multiplying every term, except those of the dimensions i and $i-1$ by a suitable power of

$$\frac{x^2}{a^2} + \frac{y^2}{b^2} + \frac{z^2}{c^2},$$

we shall express the density by the sum of two rational integral functions of x, y, z of the degrees i, $i-1$, respectively. The number of terms in these will be

$$\frac{(i+1)(i+2)}{2} + \frac{i\,(i+1)}{2} \text{ or } (i+1)^2.$$

And any ellipsoidal surface harmonic of the degree i, $i-2$... in x, y, z, may, by suitably introducing the factor

$$\frac{x^2}{a^2}+\frac{y^2}{b^2}+\frac{z^2}{c^2},$$

be expressed as a homogeneous function of x, y, z of the degree i; also any such harmonics of the degree $i-1$, $i-3$... in x, y, z may be similarly expressed as a homogeneous function of x, y, z of the degree $i-1$. And the total number of these expressions will, as just shewn, be $(i+1)^2$, hence by assigning to them suitable coefficients, any distribution of density according to a rational integral function of x, y, z may be expressed by a series of surface ellipsoidal harmonics, and the potential at any internal or external point by the corresponding series of solid ellipsoidal harmonics.

33. Since any function of the co-ordinates of a point on the surface of a sphere may be expressed by means of a series of surface spherical harmonics, we may anticipate that any function of the elliptic co-ordinates v, v' may be expressed by a series of surface ellipsoidal harmonics. No general proof, however, appears yet to have been given of this proposition. But, assuming such a development to be possible at all, it may be shewn, by the aid of the proposition proved in Art. 15 of this Chapter, that it is possible in only one way, in exactly the same way as the corresponding proposition for a spherical surface is proved in Chap. IV. Art. 11.

The development may then be effected as follows. Denoting the several surface harmonics of the degree i in x, y, z, or $\frac{i}{2}$ in v, v', by the symbols $V_i^{(1)}$, $V_i^{(2)}$, ... $V_i^{(2i+1)}$, and by $F(v, v')$ the expression to be developed, assume

$$F(v, v') = C_0V_0 + C_1^{(1)}V_1^{(1)} + C_1^{(2)}V_1^{(2)} + C_1^{(3)}V_1^{(3)} + \dots$$
$$+ C_i^{(1)}V_i^{(1)} + \dots + C_i^{(\sigma)}V_i^{(\sigma)} + \dots$$

Then multiplying by $eV_i^{(\sigma)}$ and integrating all over the surface, we have

$$\int eF(v, v')\, V_i^{(\sigma)}\, dS = C_i^{(\sigma)} \int e\,(V_i^{(\sigma)})^2\, dS.$$

The values of $\int eF(v, v')V_i^{(\sigma)}\,dS$, and of $\int e\,(V_i^{(\sigma)})^2\,dS$ must be ascertained by introducing the rectangular co-ordinates x, y, z, or in any other way which may be suitable for the particular case. The coefficients denoted by C are thus determined, and the development effected.

EXAMPLES.

1. Prove that $(\sin\theta)^4 = \frac{8}{15}P_0 - \frac{16}{21}P_2 + \frac{8}{35}P_4$.

Why cannot $(\sin\theta)^3$ be expanded in a finite series of spherical harmonics?

2. Prove that $1 + \frac{1}{2}P_1 + \frac{1}{3}P_2 + \frac{1}{4}P_3 + \ldots = \log\dfrac{1+\sin\frac{\theta}{2}}{\sin\frac{\theta}{2}}$.

3. Establish the equations

$$(\mu^2-1)\frac{dP_n}{d\mu} = n\mu P_n - nP_{n-1},$$

$$nP_n = (2n-1)\mu P_{n-1} - (n-1)P_{n-1}.$$

4. If $\mu = \cos\theta$, prove that

$$P_i(\mu) = 1 - i(i+1)\sin^2\frac{\theta}{2} + \ldots + (-1)^m \frac{\underline{|i+m}}{(\underline{m})^2\,\underline{|i-m}}\left(\sin^2\frac{\theta}{2}\right)^m + \ldots$$

and also that

$$P_i(\mu) = (-1)^i + (-1)^{i+1} i(i+1)\cos^2\frac{\theta}{2} + \ldots$$

$$+ (-1)^{i+m}\frac{\underline{|i+m}}{(\underline{|m})^2\,\underline{|i-m}}\left(\cos^2\frac{\theta}{2}\right) + \ldots$$

5. Prove that, if a be greater than c, and i any odd integer greater than m,

$$\int_{-1}^{1}(a^2 - 2ac.\mu + c^2)^{-\frac{2m+1}{2}}d\mu = \frac{1}{a^{m+1}c^m}\,\frac{2}{\underline{|2m}}\,\Sigma\,\frac{\underline{|i+m}}{\underline{|i-m}}\,\frac{c^i}{a^i}.$$

6. Prove that $\int_{-1}^{1}\left(\frac{dP_i}{d\mu}\right)^2 d\mu = i(i+1)$.

7. Prove that, when $\mu = \pm 1$, $\frac{d^m P_i}{d\mu^m} = \frac{\lfloor i+m}{\lfloor i-m} \frac{(\pm 1)^{m+i}}{2^m \lfloor m}$.

8. Prove that $\begin{vmatrix} 1, & P_1 & \dots & P_i \\ P_1, & P_2 & \dots & P_{i+1} \\ \dots & \dots & \dots & \dots \\ P_i, & P_{i+1} & \dots & P_{2i-1} \end{vmatrix}$ is a numerical multiple of $(\mu^2-1)^{\frac{i(i-1)}{2}}$.

9. Prove the following equation, giving any Laplace's coefficient in terms of the preceding one:

$$P_{n+1} = pP_n + n\int_0^p P_n dp + C,$$

where $Cp = \mu\mu' + \sqrt{1-\mu^2}\sqrt{1-\mu'^2}\cos(\omega - \omega')$ and C is zero if n be even, and

$$(-1)^{\frac{n+1}{2}} \frac{\lfloor n+1}{2^{n+1}\{\lfloor \tfrac{1}{2}(n+1)\}^2}, \text{ if } n \text{ be odd.}$$

10. If i, j, k be three positive integers whose sum is even, prove that

$$\int_1^1 P_i P_j P_k d\mu$$
$$= \frac{1.3\dots(j+k-i-1)}{2.4\dots(j+k-i)} \frac{1.3\dots(k+i-j-1)}{2.4\dots(k+i-j)} \frac{1.3\dots(i+j-k-1)}{2.4\dots(i+j-k)}$$
$$\frac{2.4\dots(i+j+k)}{1.3\dots(i+j+k-1)} \frac{1}{i+j+k+1}.$$

Hence deduce the expansion of $P_i P_j$ in a series of zonal harmonics.

11. Express $x^2y + y^3 + yz + y + z$ as a sum of spherical harmonics.

12. Find all the independent symmetrical complete harmonics of the third degree and of the fifth negative degree.

13. Matter is distributed in an indefinitely thin stratum over the surface of a sphere whose radius is unity, in such a manner that the quantity of matter laid on an element (δS) of the surface is
$$\delta S\,(1 + ax + by + cz + fx^2 + gy^2 + hz^2),$$

where x, y, z are rectangular co-ordinates of the element δS referred to the centre as origin, and a, b, c, f, g, h are constants. Find the value of the potential at any point, whether internal or external.

14. If the radius of a sphere be r, and its law of density be $\rho = ax + by + cz$, where the origin is at the centre, prove that its potential at an external point (ξ, η, ζ) is $\frac{4\pi r^5}{15R^3}(a\xi + b\eta + c\zeta)$ where R is the distance of (ξ, η, ζ) from the origin.

15. Let a spherical portion of an infinite quiescent liquid be separated from the liquid round it by an infinitely thin flexible membrane, and let this membrane be suddenly set in motion, every part of it in the direction of the radius and with velocity equal to S_i, a harmonic function of position on the surface. Find the velocity produced at any external or internal point of the liquid. State the corresponding proposition in the theory of Attraction.

16. Two circular rings of fine wire, whose masses are M and M', and radii a and a', are placed with their centres at distances b, b', from the origin. The lines joining the origin with the centres are perpendicular to the planes of the rings, and are inclined to one another at an angle θ. Shew that the potential of the one ring on the other is

$$MM' \Sigma_{n=0}^{n=\infty} \left(\frac{1}{C^{m+1}} B_n B'_n Q_n \right),$$

where $B_n = b^n - \frac{n(n-1)}{2 \cdot 2} b^{n-2}a^2 + \frac{n(n-1)(n-2)(n-3)}{2 \cdot 2 \cdot 4 \cdot 4} b^{n-4}a^4 - \ldots$

and B'_n and Q_n are the same functions of b' and a' and of $\cos\theta$ and $\sin\theta$ respectively, and c is the greater of the two quantities $\sqrt{a^2 + b^2}$ and $\sqrt{a'^2 + b'^2}$.

17. A uniform circular wire, of radius a, charged with electricity of line-density e, surrounds an uninsulated concentric spherical conductor of radius c; prove that the electrical density at any point of the surface of the conductor is

$$-\frac{e}{2c}\left(1 - 5 \cdot \frac{1}{2} \frac{c^2}{a^2} P_2 + 9 \cdot \frac{1 \cdot 3}{2 \cdot 4} \frac{c^4}{a^4} P_4 - 13 \cdot \frac{1 \cdot 3 \cdot 5}{2 \cdot 4 \cdot 6} \frac{c^6}{a^6} P^6 + \ldots\right),$$

the pole of the plane of the wire being the pole of the harmonics.

18. Of two spherical conductors, one entirely surrounds the other. The inner has a given potential, the outer is at the potential zero. The distance between their centres being so small that its square may be neglected, shew how to find the potential at any point between the spheres.

19. If the equation of the bounding surface of a homogeneous spheroid of ellipticity ϵ be of the form

$$r = a\left(1 - \frac{2}{3}\epsilon P_2\right),$$

prove that the potential at any external point will be

$$\frac{M}{r} - \frac{C-A}{r^3} P_2,$$

where C and A are the equatoreal and polar moments of inertia of the body.

Hence prove that V will have the same value if the spheroid be heterogeneous, the surfaces of equal density differing from spheres by a harmonic of the second order.

20. The equation $R = a(1 + \alpha y)$ is that of the bounding surface of a homogeneous body, density unity, differing slightly in form and magnitude from a sphere of radius a; α is a small quantity the powers of which above the second may be neglected; and y is a function of two co-ordinate angles, such that

$$y = Y_0 + Y_1 + \dots + Y_n + \dots, \quad y^2 = Z_0 + Z_1 + \dots + Z_n \dots$$

where Y_0, Y_1 ... Z_0, Z_1 ... are Laplace's functions. Prove that the potential of the body's attraction on an external particle, the distance of which from the origin of co-ordinates is r, is given by the equation

$$V = \frac{4\pi^3 a^3}{3r} + \frac{4\pi \alpha a^3}{r}\left\{Y_0 + \frac{a}{3r} Y_1 + \dots + \frac{a^n}{(2n+1)r^n} Y_n + \dots\right\}$$

$$+ \frac{4\pi\alpha^2 a^3}{r}\left\{Z_0 + \frac{a}{2r} Z_1 + \dots + \frac{n+2}{4n+2}\frac{a^n}{r^n} Z_n + \dots\right\}.$$

21. If M be the mass of a uniform hemispherical shell of radius c, prove that its potential, at any point distant r from the centre, will be

$$\frac{M}{2c} + \frac{1}{2}\frac{M}{c^2}\left(\frac{1}{2}P_1 r - \frac{1}{2.4}P_3\frac{r^3}{c^2} + \frac{3}{2.4.6}P_5\frac{r^5}{c^4} - \frac{3.5}{2.4.6.8}P_7\frac{r^7}{c^6} + \dots\right),$$

or $$\frac{M}{2r} + \frac{1}{2}M\left(\frac{1}{2}P_1\frac{c}{r^2} - \frac{1}{2.4}P_3\frac{c^3}{r^4} + \frac{3}{2.4.6}P_5\frac{c^5}{r^6} - \frac{3.5}{2.4.6.8}P_7\frac{c^7}{r^8} + \dots\right),$$

according as r is less or greater than c; the vertex of the hemisphere being at the point at which $\mu = 1$.

22. A solid is bounded by the plane of xy, and extends to infinity in all directions on the positive side of that plane. Every point within the circle $x^2 + y^2 = a^2$, $z = 0$ is maintained at the uniform temperature unity, and every point of the plane xy without this circle at the uniform temperature 0. Prove that, when the temperature of the solid has become permanent, its value at a point distant r from the origin, and the line joining which to the origin is inclined at an angle θ to the axis of z will be

$$P_0 - P_1\frac{r}{a} + \frac{1}{2}P_3\frac{r^3}{a^3} - \frac{1.3}{2.4}P_5\frac{r^5}{a^5} + \dots - (-1)^i\frac{1.3\dots(2i-1)}{2.4\dots 2i}P_{2i+1}\frac{r^{2i+1}}{a^{2i+1}} + \dots$$

if $r < a$, and

$$\frac{1}{2}P_1\frac{a^2}{r^2} - \frac{1.3}{2.4}P_3\frac{a^4}{r^4} + \dots - (-1)^i\frac{1.3\dots(2i-1)}{2.4\dots 2i}P_{2i+1}\frac{a^{2i}}{r^{2i}} + \dots$$

if $r > a$.

23. Prove that the potential of a circular ring of radius c, whose density at any point is $\cos m\psi$, $c\psi$ being the distance of the point measured along the ring from some fixed point, is

$$2\pi \cos m\phi\, (\sin\theta)^m \left\{\frac{1}{2.4.6\ldots 2m}\frac{d^m P_m}{d\mu^m}\frac{c^m}{r^{m+1}} + \frac{1}{2.4.6\ldots(2m+2)}\frac{d^m P_{m+2}}{d\mu^m}\frac{c^{m+2}}{r^{m+3}} + \ldots + \frac{1.3.5\ldots(2k-1)}{2.4.6\ldots 2(m+k}\frac{d^m P_{m+2k}}{d\mu^m}\frac{c^{m+2k}}{r^{m+2k+1}} + \ldots\right\},$$

where r is greater than c. If r be less than c, r and c must be interchanged.

24. A solid is bounded by two confocal ellipsoidal surfaces, and its density at any point P varies as the square on the perpendicular from the centre on the tangent plane to the confocal ellipsoid passing through P. Prove that the resultant attraction of such a solid on any point external to it or forming a part of its mass is in the direction of the normal to the confocal ellipsoid passing through that point, and that the solid exercises no attraction on a point within its inner surface.

CAMBRIDGE: PRINTED BY C. J. CLAY, M.A. AT THE UNIVERSITY PRESS.